Environmental Physics

Why might global warming cause increased rain and storms? Why is environmental degradation so often irreversible? Why has the problem of nuclear waste disposal still not been resolved? These and other controversial environmental issues revolve around complex scientific arguments, which can be better understood with a grasp of the key scientific concepts. *Environmental Physics* provides an introduction to the physical principles that underlie environmental issues and shows how they contribute to the interdisciplinary field of environmental science.

The book explores a broad range of topics, encompassing the natural and human environments. These include:

- natural processes – global climate and the greenhouse effect, ozone depletion, the Earth's structure and properties, hydrology, pollutant transport and biophysics;
- environmental technologies – flue-gas clean-up, noise pollution, remote sensing, renewable energy production, radioactive waste management and spectroscopic analysis;
- physical concepts – mechanics, energy, thermodynamics, electromagnetic radiation, atomic spectra, fluid flow, atmospheric processes, sound waves and radioactivity.

These fundamental and applied concepts of physics are presented in a clear, easily comprehensible fashion, made relevant by their application to topical environmental issues.

Environmental Physics makes the subject accessible to those with little previous knowledge of physics. As a student of environmental science, the reader will find the wide range of topics covered in this single volume invaluable. *Environmental Physics* is highly illustrated with over 100 figures and plates, and has boxed case studies, end of chapter summaries, further reading and a glossary.

Clare Smith was formerly a Lecturer at Imperial College London and the University of Sunderland, and is now a freelance Environmental Consultant.

Routledge Introductions to Environment Series
Published and Forthcoming Titles

Titles under Series Editors:
Rita Gardner and A.M. Mannion

Environmental Science texts

Atmospheric Processes and Systems
Natural Environmental Change
Biodiversity and Conservation
Ecosystems
Environmental Biology
Using Statistics to Understand the
  Environment
Coastal Systems

Forthcoming:
Environmental Chemistry (December 2001)

Titles under Series Editor:
David Pepper

Environment and Society texts

Environment and Philosophy
Environment and Social Theory
Energy, Society and Environment
Environment and Tourism
Gender and Environment
Environment and Business

Environment and Politics, 2nd edition (July 2001)

Routledge Introductions to Environment

Environmental Physics

Clare Smith

Routledge
Taylor & Francis Group

LONDON AND NEW YORK

First published 2001
by Routledge
2 Park Square, Milton Park, Abingdon, Oxon OX14 4RN

Simultaneously published in the USA and Canada
by Routledge
711 Third Avenue, New York, NY 10017

Routledge is an imprint of the Taylor & Francis Group, an informa business

Typeset in Times by Florence Production Ltd, Stoodleigh, Devon

British Library Cataloguing in Publication Data
A catalogue record for this book is available from the British Library

Library of Congress Cataloging in Publication Data
Smith, Clare, 1964–
    Environmental physics / Clare Smith.
        p. cm. – (Routledge introductions to environment series)
    Includes bibliographical references and index
    1. Physics. 2. Environmental sciences. I. Title. II. Series
    QC28.S593 2001
    530–dc21   00-054741

ISBN 13: 978-0-415-20190-2 (hbk)
ISBN 13: 978-0-415-20191-9 (pbk)

Contents

Series editors' preface
Environmental Science titles

The last few years have witnessed tremendous changes in the syllabi of environmentally-related courses at Advanced Level and in tertiary education. Moreover, there have been major alterations in the way degree and diploma courses are organised in colleges and universities. Syllabus changes reflect the increasing interest in environmental issues, their significance in a political context and their increasing relevance in everyday life. Consequently, the 'environment' has become a focus not only in courses traditionally concerned with geography, environmental science and ecology but also in agriculture, economics, politics, law, sociology, chemistry, physics, biology and philosophy. Simultaneously, changes in course organisation have occurred in order to facilitate both generalisation and specialisation; increasing flexibility within and between institutions is encouraging diversification and especially the facilitation of teaching via modularisation. The latter involves the compartmentalisation of information which is presented in short, concentrated courses that, on the one hand are self contained but which, on the other hand, are related to pre-requisite parallel, and/or advanced modules.

These innovations in curricula and their organisation have caused teachers, academics and publishers to reappraise the style and content of published works. Whilst many traditionally-styled texts dealing with a well-defined discipline, e.g. physical geography or ecology, remain apposite there is a mounting demand for short, concise and specifically-focused texts suitable for modular degree/diploma courses. In order to accommodate these needs Routledge have devised the Environment Series which comprises Environmental Science and Environmental Studies. The former broadly encompasses subject matter which pertains to the nature and operation of the environment and the latter concerns the human dimension as a dominant force within, and a recipient of, environmental processes and change. Although this distinction is made, it is purely arbitrary and is made for practical rather than theoretical purposes; it does not deny the holistic nature of the environment and its all-pervading significance. Indeed, every effort has been made by authors to refer to such interrelationships and to provide information to expedite further study.

This series is intended to fire the enthusiasm of students and their teachers/lecturers. Each text is well illustrated and numerous case studies are provided to underpin general theory. Further reading is also furnished to assist those who wish to reinforce and extend their studies. The authors, editors and publishers have made every effort to provide a series of exciting and innovative texts that will not only offer invaluable learning resources and supply a teaching manual but also act as a source of inspiration.

A. M. Mannion and Rita Gardner
1997

Series International Advisory Board

Australasia: Dr P. Curson and Dr P. Mitchell, Macquarie University

North America: Professor L. Lewis, Clark University; Professor L. Rubinoff, Trent University

Europe: Professor P. Glasbergen, University of Utrecht; Professor von Dam-Mieras, Open University, The Netherlands

Note on the text

Bold is used in the text to denote words defined in the Glossary. It is also used to denote key terms.

Plates

Figures

Tables

Boxes

Acknowledgements

I would like to thank the following for granting permission to reproduce images in this work: Gaylon S. Campbell for four diagrams reproduced from *An Introduction to Environmental Biophysics* (1977), with permission from Springer-Verlag; Dr Russell Thompson for a diagram of idealised airflow at 3 km elevation, from *Atmospheric Processes and Systems* (1998), reproduced by permission from Routledge; the ISO for their graph of equal-loudness level contours, from *Acoustics – Normal Equal-loudness Level Contours*, Report ISO 226:1987 (1987); H. van Dop, for a figure showing plume shapes from a high chimney, reproduced from *Luchtverontreiniging, Bronnen, Verspreiding, Transformatie en Depositie*, with permission from KNMI. For photographic images, Dr J. S. Griffiths of Plymouth University for his photo of a landslide; Bonus Energy A/S, Brande, Denmark for the photo of Vindeby wind farm; Professor Peter Brimblecombe, University of East Anglia for a photo of smoke rising from a stack; Dundee Satellite Receiving Station for the image of a North Atlantic depression; and Dr Nigel Houghton of Rutherford Appleton Laboratory for the three other remote sensing images, including the cover photo, sea surface temperature and deforestation in Brazil, from the Along Track Scanning Radiometer (ATSR) project (more details and images are available online at http://www.atsr.rl.ac.uk/), copyright CLRC/NERC/BNSC/ESA where those organisations are Central Laboratory for the Research Council, Natural Environment Research Council, British National Space Centre and European Space Agency. Most of the diagrams were produced using CorelDRAW® 9, and include clipart images from that package, which are protected by the copyright laws of the US, Canada and elsewhere. Used under licence.

Every effort has been made to contact copyright holders for their permission to reprint material in this book. The publishers would be grateful to hear from any copyright holder who is not here acknowledged and will undertake to rectify any errors or omissions in future editions of this book.

Finally I would like to thank John Maskall for all his help in the preparation of this book, and Daisy and Chloe for distracting me and putting up with me while writing it.

The cover photo is a thermal (12 μm) image of the English Channel by night, taken on 7 September 1991, where London and other urban areas can clearly be seen because of their higher temperature. Land temperatures range from 5–15°C, while sea temperatures are higher than land, with coastal water reaching 17°C. Other features such as clouds and the River Seine are also visible due to their temperatures.

Introduction

The study of the environment requires a truly interdisciplinary approach – a holistic approach. Physics by contrast is perceived as the most reductionist of the sciences, taking every system apart, simplifying it and analysing each component's behaviour, but rarely that of the whole. More recently many physicists have adopted a more systemic attitude, recognising that you cannot observe one aspect of a system without repercussions throughout the system. This is particularly relevant to complex environmental systems, where random variations and chaos are the rule rather than the exception.

Physics forms an important plank in the building of environmental science – along with many other disciplines ranging from biology and geography through to social and political sciences. Physical laws are fundamental to natural systems and to technologies across many fields in environmental science, with numerous examples. The low efficiency of photosynthesis can be explained by the quantum energy of photons involved, while the efficiency of a power station is limited by the laws of thermodynamics. Heat flow and the properties of radiation are important in understanding Earth's climate and our influence upon it; study of fluid dynamics can be used to describe the dispersion of air and water pollutants; while the laws of nuclear physics constrain the disposal of radioactive waste.

This book provides an introduction to the physical principles that underlie environmental issues – and aims to show how they contribute to environmental science as a whole. The topics chosen are broad, covering the physics that underpins both natural processes and environmental technologies. The former include aspects of the Earth's climate, geological structures, hydrology and biophysics, while environmental technologies covered include those for flue-gas cleaning, water purification, laboratory spectroscopy and renewable energy production, amongst others. It is also hoped that the reader will gain some insights into how an environmental physicist understands the world and the insights that such understanding brings.

The text is intended to cover a broad remit at a level that makes the subject accessible to a wide audience, including many with little prior knowledge of physics. For this reason mathematical treatment is kept to a minimum and the use of calculus is avoided, although simple equations are used as the most appropriate language to describe many phenomena. The emphasis is on application rather than theoretical underpinnings, so in many cases concepts are described qualitatively rather than quantitatively. For many topics, the reader will be directed towards more detailed further reading, and other more specialised volumes in this series that expand on key themes.

1 The forces of nature

Landslides, tornadoes, falling raindrops and sedimentation are all natural systems that can be understood by Newtonian mechanics – the study of forces, momentum and motion. The following key concepts are covered in this chapter:

- **Many natural systems contain motion, forces and momentum that can be described by Newton's laws**
- **Frictional forces are important in any real-world motion, dissipating energy**
- **Gravity acts between any two bodies, dependent on their mass and distance apart**
- **Rotational motion can be described by laws and equations analogous to those for straight line motion, particularly relevant to climate and orbits**
- **Different types of wave observed in environmental systems have certain properties in common**
- **Electricity and magnetism are inextricably linked, as one induces the other**
- **The Earth's magnetic field provides a tool to investigate the geological history of the Earth and a subsurface surveying technique**

Newton's laws underpin areas such as gravity, friction and rotational dynamics, some of the key processes both in the natural environment and in a number of environmental control technologies. Many environmental systems contain waves – electromagnetic waves, sound waves, seismic waves in the Earth's interior or waves on the sea – which have certain properties in common. Electrical and magnetic fields and forces are important in several environmental applications, in particular related to the magnetic field of the Earth. This chapter deals with some of these fundamentals of physics that will be applied to a broad range of applications later in the book. Given limited space, the aim is to highlight key processes and concepts, such that the reader may seek further details elsewhere. For those who need help with rearranging of equations or basic mathematics, refer to Appendix A – Mathematical hints.

Fundamentals: Newtonian mechanics

- Why are blue whales bigger than elephants?
- When is someone going to invent a car that runs on air and water?
- Why don't we just bottle up CO_2 and shoot it into space?
- Why are skyscrapers taller than trees?

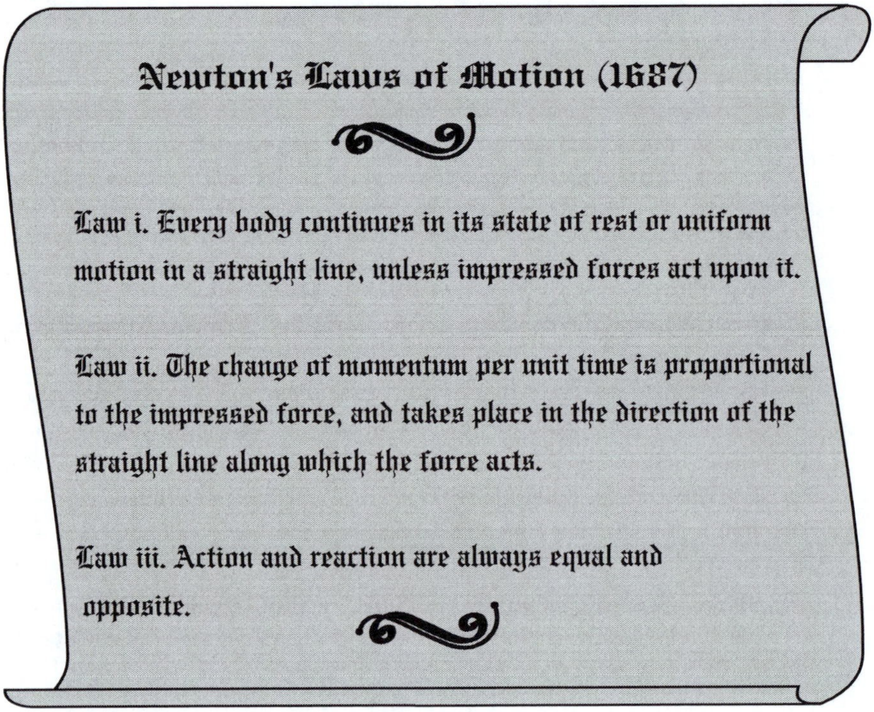

Newton's Laws of Motion (1687)

Law i. Every body continues in its state of rest or uniform motion in a straight line, unless impressed forces act upon it.

Law ii. The change of momentum per unit time is proportional to the impressed force, and takes place in the direction of the straight line along which the force acts.

Law iii. Action and reaction are always equal and opposite.

Figure 1.1 *Newton's laws of motion.*

These questions and many more can be answered using mechanics, which concerns itself with forces, momentum, energy and motion. Mechanics is built upon the work of Newton, who, back in 1687, had a vision – that the jumble of phenomena we observe in the natural world can be broken down, simplified, special cases taken and provisions made, and then described precisely by mathematical equations. While the pure Newtonian model is now known to be a simplification, and has been superseded by quantum mechanics and relativity in some areas, it still provides the building blocks that underpin many other areas of physics. This chapter will start by looking at Newton's laws of motion.

Momentum and inertia

Newton's first law is a statement of the principle of **conservation of momentum**, which is one of six fundamental conserved properties in physics. All theory and evidence shows this to be true in all cases.

Momentum, the tendency to continue moving once doing so, is defined as mass multiplied by velocity, measured in $kg\ m\ s^{-1}$ (kilogram metres per second).

Momentum is denoted u, thus $u = mv$ where m is mass and v is velocity. So the heavier something is and/or the faster it is moving, the more momentum it has.

The image of a world consisting of everything moving in straight lines without stopping is somewhat at odds with everyday experience – this is not because Newton was wrong, but because of that proviso 'unless impressed forces act upon it'. A force is anything that pushes or pulls – including air resistance, friction and gravity that are ever present.

An example of the conservation of momentum is the rebound of a gun when a bullet is fired. The bullet has forward momentum, and as the total momentum of bullet and gun is initially zero (assuming it is stationary when being fired), it must still be zero after being fired. The bullet's momentum is balanced by the gun's momentum in the opposite direction. The amount of rebound thus depends on the relative masses of gun and bullet, and the velocity of the bullet.

Momentum is of particular importance in atmospheric processes, as collisions between molecules and the exchange of their momentum control diffusion, heat flow and many fluid dynamic properties (see Chapter 4). Collisions are also relevant to a range of environmental processes, especially erosive forces such as the break-up of rocks into smaller pebbles and sand, and movements in sediments where collisions between the particles are important.

When two objects collide, momentum is conserved – because of the law of conservation of momentum. The total momentum after the collision is the same as that before, and changes in velocity may be calculated. A collision can be **elastic** or **inelastic**, or – like most real collisions – somewhere in between the two. In an elastic collision, the kinetic energy of the two objects stays the same, whereas in an inelastic collision some kinetic energy is converted into heat or sound. In elastic collisions the objects bounce off each other, while in inelastic collisions they may coalesce.

Inertia describes the same concept, the tendency for something to continue in motion or at rest unless it is pushed or pulled.

Forces

A force is simply a push or a pull, but it can be more closely defined than this. The second of Newton's laws states that rate of change of momentum is proportional to the force applied. In other words, the heavier something is and the faster it is moving, the more difficult it is to stop it or change its direction. A large heavy animal such as a horse or a wildebeest finds it more difficult to speed up and to turn than a light one such as a cheetah, which can accelerate and twist and turn easily because of its lower weight. Mathematically, the force is equal to the rate of change in momentum given by:

$$f = \Delta u \,/\, \Delta t$$

where f is the force, u is momentum, t is time and the symbol Δ means a change in the following quantity.

The definition of a force as exchange of momentum can be visualised in a jet engine. Air is drawn in at the front by a fan and ejected together with exhaust gases in a jet at the back. The momentum of the moving air is transferred to the aeroplane or boat, exerting a forwards force on it. Rocket engines work purely on this principle, ejecting exhaust gases to provide momentum, as in space there is nothing to push against.

This can be taken further, as momentum is mass times velocity. Velocity is defined as the change in distance over time. When something speeds up, the change in velocity over time is given by acceleration – if it slows down acceleration is negative. Mathematically:

$$v = \text{distance/time} = \Delta s/\Delta t$$

$$a = \text{velocity/time} = \Delta v/\Delta t$$

where s is distance, v is velocity, a is acceleration, t is time, and again Δ means a change. Now as momentum is mass times velocity, $u = mv$, the equation for a force can be given as:

$$f = \Delta(mv)/\Delta t$$

$$= m\Delta v/\Delta t$$

As mass doesn't change, the change in momentum over time is the same as the mass multiplied by the change in velocity over time. But the change of velocity over time is acceleration, so:

$$f = ma \tag{1.1}$$

This definition of a force, as mass multiplied by acceleration, is an important equation in mechanics. By rearranging this equation, if the mass and the strength of the force are known, acceleration can be calculated.

Action and reaction

The third law introduces the concept of action and reaction. If an object is lying on a table, there is a force, due to gravity on the object, exerted on the table. The reason the object does not fall through the table on to the floor is that the table is also exerting a force upwards on the object. This force of reaction is, by the third law, equal to the force of gravity exerted by the object, in the opposite direction.

A car, or a rowing boat, is propelled by this principle – a force by the tyre against the ground or the oar against the water produces a reactive force, which drives the car or boat forwards.

The three laws together describe a simple model of forces and motion. For instance consider a tennis serve. As the racket moves towards the ball it has momentum, which depends on its weight and the speed at which the player strikes the ball. At the moment of impact, the racket exerts a force on the ball (action), and the ball exerts an equal and opposite force on the racket (reaction). The ball accelerates according to its mass and the size of this force, and likewise the racket decelerates. Momentum is transferred from racket to ball due to these forces, resulting in their changes in velocity. Mathematically it all adds up – momentum is conserved overall, the force depends on how fast this momentum is transferred from racket to ball, and the force of reaction allows the racket to decrease its momentum accordingly. If you measured the speed of the racket head just before and just after the moment of impact, and the masses of the racket and the ball, you could work out how fast the serve was.

Motion

If the velocity that an object is travelling at is constant, it can be calculated from time and distance travelled using:

velocity = distance/time

or:

$v = \Delta s / \Delta t$

If the object's velocity varies, this is no longer true. For the special case of motion in a straight line when acceleration is constant (i.e. an object subject to a constant force), the distance, velocity and acceleration are linked by three simple equations. Given the conventional notation, call distance s, velocity u at the start, when time t = zero, velocity v thereafter, and acceleration a. The three equations of motion are:

$$v = u + at \tag{1.2}$$

$$s = ut + \tfrac{1}{2}at^2 \tag{1.3}$$

$$v^2 = u^2 + 2as \tag{1.4}$$

Equations (1.2) and (1.3) can be demonstrated graphically, while Equation (1.4) can be derived from the other two by simple substitution. The three equations of motion can be used to calculate distance, velocity, acceleration, or time taken if any two of these are known. Together with Equation (1.1) defining a force, $f = ma$, this provides the basis for general solutions of problems concerning forces and motion, as described in any basic physics text.

SI units

It is important in physics to use a consistent set of units – otherwise the equations will no longer hold. The equation speed = distance/time is true if speed is in metres per second, distance in metres and time in seconds. It obviously cannot be true if speed is in km/h or distance is in miles, without allowing for converting units. The SI system (Système Internationale) provides a standardised unit system, which must be stuck to, to get the right answers. An answer, or any number, is meaningless without a unit.

All SI units are derived from kilograms, metres, seconds, and coulombs (for electric charge). Very large and small quantities use the prefixes kilo, mega, micro and so on, listed in Appendix B.

In SI units, as distances are measured in metres and time in seconds, hence velocity is in metres per second, $m\ s^{-1}$ (the minus sign in the superscript means 'per'). Acceleration, the increase in velocity per second, is in metres per second per second, or $m\ s^{-2}$. The SI unit of force is the Newton, defined as $1\ N = 1\ kg\ m\ s^{-2}$, i.e. the force required to accelerate a mass of 1 kg by $1\ m\ s^{-2}$.

In any equation, the units on one side must be the same as those on the other. You cannot add apples and oranges, and likewise you cannot add metres to seconds. This gives a handy method of checking the validity of an equation – if the units are wrong, the equation is wrong. For instance, as force is given by $f = ma$ in Equation (1.1), the unit of force must be the same as that of mass multiplied by acceleration, which means kg multiplied by $m\ s^{-2}$, hence $kg\ m\ s^{-2}$. To simplify things this is called a **Newton**, denoted N.

Some quantities do not have a unit, because they are either a proportion or a **dimensionless** quantity. For instance, the porosity of stone is the proportion of air spaces to total volume. The unit is thus volume divided by volume, so it has no units – it is the same whether volumes are measured in m^3 or mm^3. Another dimensionless quantity is the **Reynolds number**, a quantity constructed to characterise fluid flow (pp. 162–3), representing a ratio of viscous to inertial forces, where viscosity is a measure of how 'treacly' the fluid is, and density is the mass of a substance per unit volume (p. 12). The Reynolds number is defined as $\rho v l/\mu$, where a fluid of density ρ (unit: $kg\ m^{-3}$) and viscosity, μ (unit: $kg\ m^{-1}\ s^{-1}$) flows at velocity v (unit: $m\ s^{-1}$) relative to a solid of length l (unit: m). The unit of the Reynolds number would thus be:

$$\text{unit} = (kg\ m^{-3}\ m\ s^{-1}\ m)/kg\ m^{-1}\ s^{-1}$$

$$= kg\ m^{-1}\ s^{-1}/kg\ m^{-1}\ s^{-1}$$

which cancels out, so it is dimensionless and has no units.

Scalars and vectors

Force, velocity and acceleration are all **vector** quantities, defined as those that have a direction associated with them as well as a magnitude. By contrast mass is a **scalar** quantity, which has magnitude but no direction. Other examples of scalar quantities could include time, the size of an object, or electric charge.

The key difference is that scalars can be added up directly (1 kg of lemons plus 2 kg of lemons makes 3 kg), while for vectors this does not apply, as the direction must also be taken into account. If a cyclist is travelling at 20 km h^{-1}, and subjected to a wind from the side at 15 km h^{-1}, adding the two numbers to 35 km h^{-1} is meaningless. To find the apparent wind, the two velocity vectors may be added either graphically as shown in Figure 1.2 (a), or algebraically. The cyclist will be subject to a wind of 25 km h^{-1}, coming at an angle. Likewise for forces, for instance a kite is subject to a horizontal force due to the wind and a vertical aerodynamic lift force, reduced by the downwards pull of gravity. This results in a net force along the direction of the kite string, with its equal but opposite reaction given by the person holding the string, as in Figure 1.2(b). The converse is that

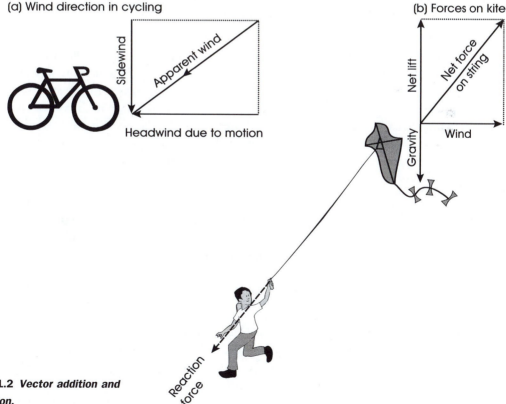

(a) Wind direction in cycling

Sidewind

Apparent wind

Headwind due to motion

(b) Forces on kite

Net lift

Net force on string

Gravity

Wind

Reaction force

Figure 1.2 *Vector addition and resolution.*

forces can be resolved back into their component parts. If the force and angle of the kite string are known, they could be used to calculate the wind speed and the lift force at the kite.

Friction and air resistance

Why do objects in common experience generally not continue moving in straight lines forever, as Newton's first law predicts they should? The answer lies in friction and air resistance. Friction is a force between two objects due to roughness of their touching surfaces, reducing their motion relative to one another, while air resistance results from frictional forces between an object and the air, slowing down any movement through it.

Friction has two forms – static friction and dynamic (or kinetic) friction. Static friction is when the two objects are still with respect to one another, and results in a larger frictional force than dynamic friction, when they are moving. Any moving object is subject to frictional forces between its moving parts, tending to slow it down, and creating heat in the process. Friction can be useful, for instance rock types used for road surfacing materials are chosen for their friction coefficients, with particular grades of stone with the highest friction coefficients sometimes being used near pedestrian crossings, allowing vehicles to stop safely in less time.

The degree of friction depends principally on the roughness of the surfaces, but will be affected by lubricants such as oil or water that flow into the tiny gaps in between solid surfaces allowing them to move more easily. In other words for a wet surface the frictional force is less, so for instance a vehicle may skid on a wet road when it brakes – the reaction force of friction is less than the force created by its rate of change of momentum in slowing down. Weight or pressure can increase friction, pressing two objects more tightly together and increasing their cohesion.

The force due to air resistance is known as a **drag** force, and is of great importance in aerodynamic design of vehicles (and birds), and in terminal velocity (pp. 17–18). The drag force from wind on natural surfaces such as leaves, water, and rock is relevant to processes such as wind pollination, resistance of plants to wind stress, and erosion. Drag depends on the object's shape, size, surface characteristics and speed, and will be least for a smooth, aerodynamic object with a relatively small surface area, moving slowly.

Drag forces can take two forms depending on flow characteristics (which will be considered in more detail on pp. 161–3). For a small object moving slowly, such as an airborne pollution particulate, or for movement in a viscous fluid (like oil or treacle) **skin drag** predominates, due to viscous forces between fluid particles around the object's surface. For faster moving objects in less viscous fluids, such as large objects in air, **form drag** is more important. This is to do with the change in momentum of the fluid as it moves out of the way, which is dependent on the object's shape.

In the nineteenth century, Stokes showed that the force on a sphere due to skin drag is:

$$F_d = 6\pi r\mu v \qquad (1.5)$$

where F_d is drag force, r is the radius of the sphere, μ is viscosity and v is velocity.

For form drag, the equivalent force is given by:

$$F_d = \tfrac{1}{2}\,\rho_a C\pi r^2 v^2 \qquad (1.6)$$

where ρ_a is the density of air (1.2 kg m^{-3}) and C is a drag coefficient that depends on the shape of the object, with the value 1.2 for a sphere. C will be highest for a flat plate perpendicular to the movement, and lowest for an aerodynamic 'teardrop' shape. Note that this force increases proportionally to the radius squared, i.e. it depends upon the surface area of the object opposed to the motion, and to velocity squared, so it predominates over skin drag at high velocities.

Gravity

Earthbound creatures live with gravity constantly, and it affects many natural environmental processes, together with being useful in certain environmental technologies and in understanding the Earth through gravity surveying. The gravitational fields of the Moon and the Sun also control the Earth's tides, which occur in the atmosphere and in the Earth's crust in addition to the better known ones in the ocean.

Newtonian gravity

While sitting under his famous apple tree, Newton postulated that gravity is an attraction between masses, and that anything falling under gravity accelerates at the same rate, no matter how heavy it is. Motion under gravity is governed by gravitational acceleration, which at the Earth's surface has the value $g = 9.8$ m s^{-2}.

The force due to gravity, f, can be given by $f = mg$, which is just Equation (1.1) with acceleration equal to g. In other words the force increases in proportion to the mass, so the acceleration remains the same for any mass.

How fast do things fall? For an apple dropping from a tree 4.5 m high into Newton's lap sitting below, 0.5 m high, the distance fallen is 4 m at an acceleration of 9.8 m s^{-2}. Putting $s = 4$, $a = 9.8$, $u = 0$ into Equation (1.2) gives:

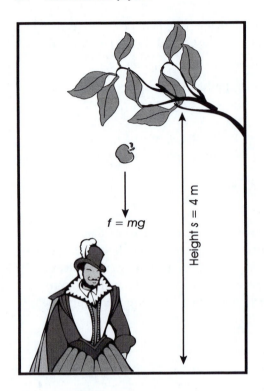

f = mg

Height s = 4 m

Figure 1.3 *Newton's apple.*

$$s = (0) + \tfrac{1}{2}at^2$$

rearranging:

$$t = \sqrt{(2s/a)}$$
$$= \sqrt{(2 \times 4/9.8)}$$
$$= 0.9 \text{ seconds}$$

The velocity of the apple would be:

$$v = u + at$$
$$= 0 + 9.8 \times 0.9$$
$$= 8.9 \text{ m s}^{-1}$$

Mass, weight and density

Mass can be seen as 'reluctance to be moved by a force'. While mass is an intrinsic quality of matter, the weight of an object is the force on it due to gravity. Weight is therefore measured in Newtons, not kg. In outer space, an object will have weight of zero, but it will have the same mass as on the Earth, and the momentum associated with it.

On the Earth, weight w is given by Equation (1.1), $f = ma$:

$$w = mg$$

So a 10 kg mass weighs $10 \times 9.8 = 98$ N.

On the moon, gravity is about one-sixth of that on the Earth. So moon gravity could be called $g_m = 9.8/6 = 1.6$ m s^{-2}. The weight of the same 10 kg mass is now:

$$w = mg_m$$
$$= 16 \text{ N}$$

The mass of an object stays the same wherever in the Universe it is.

The **density** of a substance indicates how heavy a given sized quantity is. Density is defined as mass per unit volume, measured in kg m^{-3} (kg per cubic metre). It is usually denoted by the Greek letter ρ (rho), given by $\rho = M/V$.

The density of water is 1,000 kg m^{-3}. **Relative density** of a substance means its density relative to water. For instance, the density of steel is 8,500 kg m^{-3}, so its relative density is 8.5. Any substance with relative density less than 1 will float on water.

Box 1.1

Landslides and slope stability

Gravity has been termed 'the great leveller', dragging downwards on everything at the surface of the Earth. Erosive agents such as wind, water and glaciers help greatly in moving large amounts of material, but even without these, considerable mass movements take place. These vary from minor soil slippage that may barely disturb vegetation, to major mudslides, rockslides and avalanches that can destroy large areas of crops or forest and any roads or villages in their path.

A landslide will occur if the force of gravity down a slope is greater than the frictional forces that support the soil, or in the case of rocks the shear strength of the rock involved (see Chapter 4). Because dynamic friction is generally less than static friction, once the slope starts to move it is likely to continue to do so and to accelerate, until it reaches the bottom of the slope.

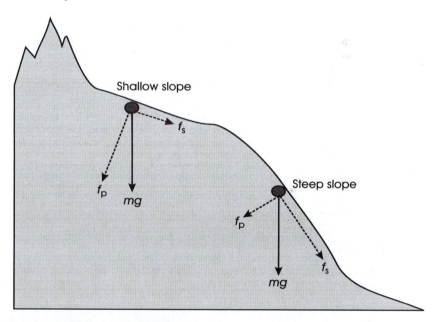

Figure 1.4 *Forces acting in landslides.*

Factors that affect the likelihood of landslides include the angle of the slope, weight of overlying material or buildings, and anything that affects frictional forces. The gravitational force on a parcel of soil, *mg*, can be resolved into components f_s parallel to the slope and f_p perpendicular to the slope (Figure 1.4). Slope is important because the steeper the angle, the greater the component of gravitational force acting down the slope, f_s. Thus all other things being equal, the steeper the slope, the more likely a landslide is to occur. The maximum incline that a specific material can rest at and still be stable is known as the **angle of repose**. Angle of repose depends principally upon the material – different soil types will vary in the frictional forces between particles, as the size and

continued

shape of the soil particles varies, and also the level of compaction of the soil. Spoil tips must be constructed with these factors in mind, at a maximum slope angle depending on the material concerned. In rocks, the strength depends not only on the rock type but the angle of bedding planes and the existence of faults or weaknesses.

Water is important for several reasons. The sheer weight of water in soil increases gravitational forces, while it also generally reduces friction by lubricating the soil, especially when saturated, although very dry soil can also become weaker and less cohesive. Under extremely wet conditions, previously solid soil can become wet enough to flow like a liquid, due to water preventing friction between soil particles. Water can also cause cracks through expansion and contraction in clays, and freeze-thaw action. Changes in drainage or increases in moisture from irrigation or household drains can therefore be causes of landslides. Perhaps more commonly very heavy rainfall can saturate a slope that was stable under normal conditions, causing sudden large and rapid movements that can be devastating. This was the trigger for the disastrous collapse of unstable minewaste heaps in Aberfan in Wales, in the 1960s, engulfing the village school and causing the deaths of dozens of children.

Vegetation plays a part, as plant and tree roots hold together soil (like reinforcing rods in concrete), and increase its capacity to hold water. Vegetation of spoil tips and tree planting in roadside cuttings can thus not only improve their appearance but also their stability. Deforestation in mountainous areas can cause landslides because of the removal of this reinforcement, making stabilisation difficult. In November 1998, Hurricane Mitch struck Honduras and Nicaragua with torrential rain and wind, creating disastrous mudslides that caused the loss of thousands of lives and destruction of the homes and livelihoods of hundreds of thousands. This natural disaster was believed to have been greatly exacerbated by deforestation of the mountainous areas of these countries over the previous years, destabilising soil and creating the preconditions for major erosion and collapse.

Plate 1.1 *Landslide on cliff top at Lyme Regis, UK.*
Source: photo by courtesy of Dr J. S. Griffiths, University of Plymouth

In general solids are more dense than their liquid form. This does not always hold however – water is unusual in that it expands (i.e. becomes less dense) when it freezes. Ice has a relative density of 0.9, and so it floats on water.

Universal gravity – big *G* and little *g*

Newton's gravitational acceleration g ('little g') is only constant at or near the Earth's surface. For a more generalised case, gravity is a force between any two objects, depending on their masses and the distance between them. The force increases proportionally to the two masses attracting, and inversely proportionally to the square of distance between them.

$$f = \frac{GMm}{R^2} \tag{1.7}$$

where f = force, M and m are the two masses, R is the distance between them, and G ('big G') is the universal gravitational constant, with the value $G = 6.67 \times 10^{-11}$ N m² kg⁻².

Hence as you travel away from the Earth the force becomes weaker. If you imagine a gravitational field radiating out from the Earth like rays from the Sun, the strength depends on how far apart the rays are. At a distance R the rays will have spread out to cover a large sphere with an area proportional to R^2, so the force decreases accordingly. This is known as an **inverse-square law**, and there are similar examples from several different branches of physics, including electrical forces and the dissipation of sound.

At the Earth's surface this same force is given by $f = mg$, and so the relation between big G and little g can be shown:

$$g = \frac{GM}{R^2} \tag{1.8}$$

where M is the Earth's mass and R its radius. As the Earth's radius is about 6,400 km, its mass can be calculated from this equation:

$$M = \frac{gR^2}{G}$$

$$= \frac{9.8 \times (6.4 \times 10^6)^2}{6.67 \times 10^{-11}}$$

$$= 6 \times 10^{24} \text{ kg}$$

Gravity is a very weak force, so it is only noticeable from very large masses such as the Earth. Although in theory any two masses attract, such as two apples, the force would be so tiny it is virtually immeasurable.

Box 1.2

Gravitational anomalies and surveying

One method of studying the interior of the Earth is by measuring very small variations in the strength of gravity at the Earth's surface. These variations in gravity are termed **gravitational anomalies**.

If you measure the constant g in an aeroplane, it will be lower than at the Earth's surface – because you are further away from Earth's centre of gravity, and gravity decreases with distance (the inverse-square law). For this reason, gravitational anomalies are corrected for the height above sea level.

If you happen to be flying over the top of Mount Everest when you measure g, you would expect it to be slightly higher than its average, corrected value for the height. This is because of the additional mass of the mountain below you. Gravity measurements can thus also be adjusted for topography – the additional mass of visible features such as mountains.

However, when gravity measurements are made at the tops of mountains, and corrected for topography, they are found to be lower than normal – they show a negative anomaly.

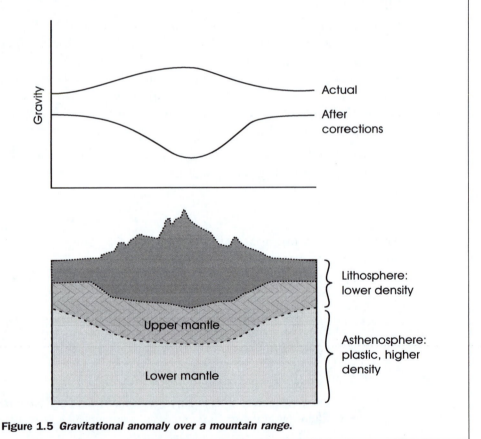

Figure 1.5 *Gravitational anomaly over a mountain range.*

This is because of features below the surface: below a mountain range lies an area of thicker crust, which has a lower density than the mantle rocks below. Thus this crustal rock 'root' has lower mass, reducing gravity very slightly. This is due to **isostatic compensation** (see Ritter 1986 pp. 38–41), caused essentially by the weight of the mountains gradually pressing the crust and solid outer layer of the mantle down, as if they were floating on the plastic asthenosphere below (Figure 1.5; see also Box 4.1).

In a similar manner, gravitational surveying can be used to explore other natural features below the Earth's surface – the thickness of the crust, areas of rocks with higher density, or plumes of denser material upwelling from the lower mantle. They can also be used as an aid to mineral prospectors and hydrographers, to explore subsurface features that may indicate the presence of ore deposits, water or other resources. Man-made features can also be studied, for instance gravitational surveys have been used to assess the extent of old landfill sites and mineworkings, where historical records are inadequate (Sharma 1997).

Terminal velocity and settling velocity

The idea of constant gravitational acceleration is true in 'idealised' circumstances, but in real life air resistance plays a part. A feather and a brick will not fall at the same rate, but each will reach a maximum speed known as their **terminal velocity**. Terminal velocity is relevant to many environmental applications involving gravity, affecting falling rain, animal characteristics, and settling of air pollution particulates. In water, it is more often known as the settling velocity, important in water treatment and sedimentation in rivers and lakes.

When falling, an object will accelerate under gravity until the drag force is equal but opposite to the force of gravity, when speed will reach its maximum and remain constant. For a human being (without a parachute), terminal velocity is about $65 \, \mathrm{m \, s^{-1}}$ ($230 \, \mathrm{km \, h^{-1}}$), reached after about 10 seconds, when falling from around 300 m. Smaller creatures such as spiders have relatively much greater surface area compared to their weight than humans, and so have relatively higher drag and much lower terminal velocity. Hence they reach terminal velocity very quickly (and also have less momentum), and can withstand falls from great heights without damage.

Higher terminal velocity means large particles settle more rapidly, leading to the kind of sorting by size seen in pebbles on the beach, or in estuarine sediments, which affect the composition of sedimentary rocks. In addition, denser particles also settle rapidly, so for instance in developing countries where lead is still commonly used as a petrol additive, lead-rich dust from vehicle exhausts may be found closer to the roadside than lighter road dust or hydrocarbon particulates of the same size.

An expression for terminal velocity can be derived from the equilibrium between the force of gravity and the drag forces. For a small sphere where viscous forces predominate (such as a dust particulate settling in air), the skin drag is given by

Equation (1.5). Equating this to the force due to gravity gives:

$$mg = 6\pi r \mu v$$

For mass m given by the volume of a sphere $(4/3\pi r^3)$ times density ρ,

$$\tfrac{4}{3}\pi r^3 \rho_s g = 6\pi r \mu v$$

$$v_t = \frac{2g\rho_s r^2}{9\mu} \tag{1.9}$$

where v_t is terminal velocity, g is acceleration due to gravity; ρ_s is the density of the sphere; μ is viscosity of the fluid; r is the radius of the sphere. This is known as Stokes' law, illustrating that settling velocity is highest for a larger, dense object in a less viscous medium.

For larger objects and more rapid flow a different expression applies, as drag forces are now described by Equation (1.6). Equating with gravitational force as before gives:

$$\tfrac{4}{3}\pi r^3 \rho_s g = \tfrac{1}{2}\rho_a C \pi r^2 v^2$$

$$v_t = \sqrt{\frac{8\rho_a r g}{3\rho_a C}} \tag{1.10}$$

where ρ_a is the density of air and C is drag coefficient. This equation can be used to describe the speed of falling raindrops. Entering numerical values into (1.10) gives the terminal velocity of a raindrop as $v_t \approx 4.3\sqrt{r}$, where r is the radius of the drop in mm. The largest drops of around 6 mm diameter fall at 7.5 m s^{-1}, while drizzle with droplets of 0.5 mm diameter falls at 2 m s^{-1} or slower. Thus a raindrop falling from a cloud at a height of 1,500 m could take from 200 seconds ($3\tfrac{1}{2}$ minutes) to 780 seconds (13 minutes) to reach the ground. Smaller droplets may also be lifted back up in updraughts, until they have grown to a size where their terminal velocity is higher than the speed of the updraught. It is evident that the finer rain will have more time as well as a larger surface area to absorb pollutants from the air during its fall, greatly affecting the impact of acid precipitation and related pollution problems.

Rotational dynamics and angular momentum

The physics that describes rotation and spin is useful to understand many natural systems, such as vortices in winds and water, and the relation between climate systems and the Earth's rotation. It is also used in the laboratory centrifuge, and certain pollution control technology.

Box 1.3

Settling chambers

These are among the simplest air pollution control devices, consisting of a large chamber in which flue gases slow down allowing particulates to drop out under gravity. They are used in many energy intensive industries, such as smelters or glassworks, where large coal-fired combustion plant are required. For a chamber of height h, length l, and gas velocity u, the gas will take l/u seconds to traverse the chamber, while a particle with settling velocity v_s will settle in h/v_s seconds. So any particle with diameter large enough such that $h/v_s < l/u$ will settle out, together with a proportion of smaller sized particles. Rearranging Stokes' law, Equation (1.9), gives a particle radius for 100 per cent settling of:

$$r = \sqrt{\frac{9\mu uh}{2\rho gl}}$$

From this formula the size of particulate that will be removed and the cleaning efficiency at smaller particle sizes can be calculated. For instance for flue gas flowing at 4 m s^{-1} through a chamber of size 0.5 m by 10 m, with particulate density of 1,350 kg m^{-3}, with gas viscosity at the operating temperature of 1.75×10^{-5}N s m^{-2}, the size for 100 per cent settling will be:

$$r = \sqrt{\frac{9 \times 1.75 \times 10^{-5} \times 4 \times 0.5}{2 \times 1350 \times 10 \times 10}}$$

$$= 34.1 \times 10^{-6} \text{ m or 34 } \mu\text{m}$$

Settling chambers provide a cheap method to remove large particles but are ineffective for finer particulates, requiring large areas and/or very slow gas flows. They are generally used as a primary treatment before a more expensive and efficient device such as a scrubber or electrostatic precipitator (Box 1.6).

The same principle is used in primary water treatment, to allow larger suspended solids to settle out before further treatment.

Angular momentum is the tendency for something that is spinning to continue to spin, whether it is a planet or a bicycle wheel. Like linear momentum, it is conserved – that is, unless external forces act, something that spins will continue to do so forever.

Rotational movement can be described by a series of equations analogous to those for linear motion and forces, but with distance, velocity, momentum and mass replaced by their rotational counterparts: angle turned, angular velocity, angular momentum and moment of inertia respectively. Linear forces are replaced by turning force termed **torque**. Angular velocity is given the symbol ω, and is equal to the tangential (i.e. straight line) velocity divided by the radius of the circle being rotated around.

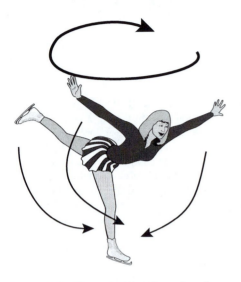

Figure 1.6 Moment of inertia and an ice-skater. As she draws her arms and legs in and down, she will spin faster.

Moments of inertia

The amount of angular momentum something has depends on how fast it is spinning, its mass, and how that mass is distributed. The distribution of that mass around the axis is represented by the moment of inertia, I. If an ice-skater is turning slowly with their arms and legs outstretched, then crouch down and hold their limbs in to their body, they will spin faster, as shown in Figure 1.6. Similarly, high divers control the rate of their spin by stretching out to slow down, curling up to spin faster.

When ice-skaters hold their limbs in to the body, they reduce their moment of inertia, as this means that the mass is closer to the axis of rotation. The increase in the rate of spin is an effect of the conservation of angular momentum. Angular momentum L is given by the product of the moment of inertia and the angular velocity:

$$L = I\omega \qquad (1.11)$$

where ω is the angular velocity. This is analogous to linear momentum, given by mass times linear velocity. As this product is constant, reducing the moment of inertia must cause an increase in angular velocity.

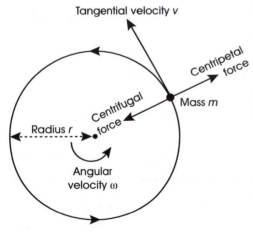

Figure 1.7 Central forces and rotation.

Central forces

When an object spins, it tends to continue in a straight line, at a tangent to the direction it is spinning, because of conservation of linear momentum. This results in a **centripetal force** pulling away from the centre of rotation. To continue spinning the centripetal force must have an equal but opposite force of reaction, pulling into the centre, which is the **centrifugal force**. If you whirl a lasso the centrifugal force is provided by the tension in the rope and the force of you holding it; in the case of the Earth going round the Sun, it is provided by gravity. The magnitude of the centripetal force is given by:

$$F_c = \frac{mv^2}{r} = mr\omega^2 \qquad (1.12)$$

where F_c is the centripetal force, m is the mass of the object, v its tangential velocity, ω its angular velocity, and r the radius of the circle moved in as shown in Figure 1.7.

The Coriolis force

Why do low pressure wind systems blow anticlockwise in the northern hemisphere but clockwise in the south? It is because of **Coriolis forces**, felt by any body moving relative to something rotating. Imagine you have a rotating turntable, and you try to roll a marble from its centre out towards the rim. Because of Newton's first law, the marble will tend to go in a straight line, but it will also be subject to the rotating motion of the turntable. The result will be a force at right angles to the marble's motion, going in the opposite direction to the rotation of the turntable. This force is the Coriolis force, and the marble will roll in a curved path as shown in Figure 1.8(a). The force increases with the marble's speed, and with the angular velocity of the turntable. The magnitude of the Coriolis force is given by:

$$f_c = 2m\omega v \qquad (1.13)$$

where m is mass of the object, ω is angular velocity and v is velocity of the object perpendicular to the rotation.

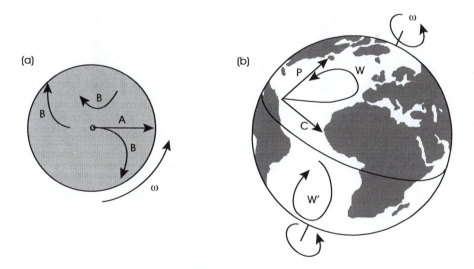

Figure 1.8 *The Coriolis force: (a) on a turntable, where A is the path of a marble on a stationary turntable, and B shows the effect of rotation ω; (b) in the atmosphere, where the air is subject to a force due to the pressure gradient, P, and the Coriolis force C, with the resulting wind patterns W and W'.*

Box 1.4

Cyclone separators for air pollution control

Emissions of particulate pollution in flue gases may be reduced using a cyclone separator. The cyclone works by directing the stream of flue gases around a conical cylinder (see Figure 1.9). Centrifugal acceleration draws the heavier particles to the outside of the cylinder, from where they will settle to the bottom and be extracted. Cleaned gas is extracted from the top, at the centre of the cylinder. The effect is like amplifying and speeding up gravitational settling, not requiring the large area and long residence times of a settling chamber (see Box 1.3).

The cleaning efficiency of the device depends on the particle size and the rotational velocity achieved. Larger, heavier particles will 'settle' to the outside wall more readily than smaller ones, due to their higher terminal velocities. Turbulence prevents very fine particles from settling, in the same way that larger stones and sand settle in a fast flowing river more readily than fine mud and clay.

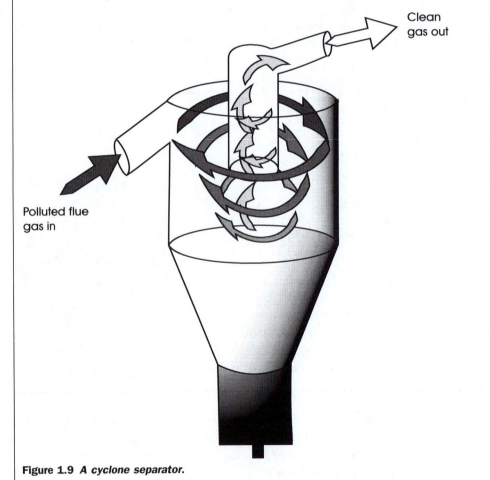

Figure 1.9 *A cyclone separator.*

An adapted form of Stokes' law (Equation (1.9)) may be used to estimate the cleaning efficiency of a cyclone, as those particles with terminal velocity high enough to reach the outer wall within a number of circuits around the cylinder will be extracted. In this case, the gravitational acceleration is replaced by the centripetal acceleration from the rotation, given by v_t^2/r, where v_t is tangential velocity and r is radius. Cleaning efficiency and the size of particulates removed can then be calculated in a similar way to in a settling chamber. The **separation factor** can be defined, which is the ratio of the terminal velocity of particles due to this centripetal force (v_r), to their normal settling velocity under gravity (v_g). Separation factor S is a dimensionless ratio, given by:

$$S = v_r/v_g = v_t^2/rg$$

Separation factors range from 5 to 2,500, indicating much more rapid settling than under gravity for particles of the same size. Using high velocities and small radii, cyclones are considerably more effective at particulate removal than settling chambers, as well as being faster and smaller. Cyclones are however still a relatively cheap, 'quick and dirty' means of flue gas cleaning. Up to 90 per cent of the larger particulates can be removed, but efficiencies are much lower for small (< 10 μm diameter) particulates.

As the Earth is rotating, anything moving on the Earth's surface at right angles to the direction of rotation is subject to Coriolis forces, including weather systems. Here the Coriolis force can be understood as a direct result of the conservation of angular momentum of the atmosphere, which spins with the Earth on its axis. A movement of air towards the axis reduces its moment of inertia, and hence the air will tend to rotate around the Earth more rapidly, as with the ice-skater. In a low pressure system in the northern hemisphere, wind blowing due north will have a component moving towards the centre of rotation of the Earth (if you imagine looking at the Earth from above the North Pole this will be obvious). Hence it will be subject to a Coriolis force in the direction of the Earth's rotation, that is eastwards. Likewise, a southerly airstream will get a westerly Coriolis force, while in the southern hemisphere the directions are reversed. These forces are small but they are sufficient to start the circular motion of weather systems.

On or near the equator no Coriolis force will be felt, as there is no motion towards or away from the centre of rotation, and so a movement does not alter the moment of inertia. It is a commonly held belief that bathwater goes down the plug anti-clockwise in the northern hemisphere and clockwise in the south. Sadly this is not the case – the direction of rotation of bathwater is due to you swirling it as you get out, or to minute irregularities in the shape of the plughole. The Coriolis force is far too small to affect it.

The vortex

Swirling bathwater is an example of a vortex, a phenomenon common in nature, observed in tornadoes, whirlpools and other climatic features. Vorticity is a measure of the rotation of a fluid about an axis. In a vortex, the rotating fluid is drawn towards the centre of rotation, but then misses and continues to rotate around it. Due to the conservation of angular momentum, as the fluid is drawn to the centre of the vortex it speeds up – as for the spinning ice-skater. Thus it moves in a spiral pattern, not actually reaching the centre. This happens in low-pressure wind systems including tornadoes, cyclonic winds in the mid-latitudes, and tropical cyclones (Box 5.1).

A small but extremely violent vortex forms, often over land, in a tornado. Here the central low pressure is caused by strong convectional updraughts from high temperatures at the surface, to colder air higher in the atmosphere. Tornadoes are generally associated with thunderstorms, and they can also occur within hurricanes. As the wind speed increases, air is sucked in and upwards in the central 'eye of the storm'. Very high wind speeds can be reached in this way, and maintained as pressure at the centre remains low. Wind speeds at the centre have been estimated

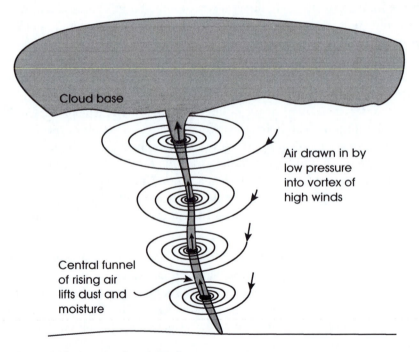

Cloud base

Air drawn in by
low pressure
into vortex of
high winds

Central funnel
of rising air
lifts dust and
moisture

Figure 1.10 *A vortex in a tornado.*

as high as 200–300 knots or around 400–600 km h^{-1} (Critchfield 1983), over a narrow area of typically a few hundred metres, which may cause considerable destruction. Tornadoes are common throughout the USA, particularly in central states in spring and early summer, but may occur virtually anywhere within the mid-latitudes and tropics, and have been known in the UK when air temperatures or windspeeds are sufficiently extreme. At sea, tornadoes cause waterspouts as sea water is sucked into the low pressure centre of the vortex.

Vortices are also present in turbulent flow, as eddies, whether in fast moving water or in the slipstream from vehicles or birds. These turbulent eddies are important in increasing drag forces, and so are aimed to be minimised in aircraft or birds' wings. The precise size and location of eddies are unpredictable, as the turbulence involves random and chaotic processes. Thus it is said that the flapping of a butterfly's wing in Siberia can create an eddy, that grows and provides the necessary initial angular momentum to set off cyclonic wind patterns, affecting weather systems across the whole of Europe. While the impact of a butterfly could never be verified, chaos theory does show that although general trends of climate can be broadly predicted, the details of when and where climatic events occur are random and therefore not theoretically predictable over the long term.

Orbits

Back in the seventeenth century, the astronomer Kepler proposed simple geometrical laws to describe the movement of the planets, which Newton developed to formulate equations. The same factors that determine planetary motion also determine the orbits used by satellites around the Earth, which are now used as bases for remote sensing of the environment. Any orbit can be described by a series of simple equations, whereby the centripetal and centrifugal forces must balance. In this case, centrifugal force comes from gravity. Near the Earth's surface, putting centripetal force from Equation (1.12) equal to gravitational force gives:

$$\frac{mv^2}{R} = mg$$

where m is the mass of the satellite, v its velocity, R is Earth's radius ($R = 6,400$ km) and g is gravity (10 m s^{-2}). Rearranging to find velocity gives:

$$v = \sqrt{(g\,R)} \tag{1.14}$$

$$= \sqrt{64,000,000}$$

$$= 8,000 \text{ m s}^{-1}$$

For a low orbit, the satellite must therefore travel at 8,000 m s^{-1}, or 8 km s^{-1}, which is 28,800 km hr^{-1}. The time period of the rotation will be given by distance travelled over velocity, or:

$$T = \frac{2\pi R}{v} \qquad (1.15)$$

$$= \frac{2\pi \times 6,400,000}{8,000}$$

$$= 5,026 \text{ seconds}$$

$$= 83 \text{ minutes}$$

At this speed, the satellite will orbit the Earth in about 83 minutes. Low orbit satellites are useful for a range of purposes including meteorological measurements and remote sensing (see Box 1.5).

If a satellite travels faster than 8 km s^{-1}, it will go into an elliptical orbit (like a comet round the Sun), and if its speed is over 11 km s^{-1}, it will reach escape velocity, where it is not in an orbit at all but flies away from the Earth into space, or more probably into the Sun.

At velocities of less than 8 km s^{-1} satellites will orbit at a higher altitude. As height increases, Earth's gravity declines, thus lower speeds are needed to balance it. Of particular use are geostationary orbits, where the satellite remains in the same position relative to the Earth's surface. For a geostationary orbit, the speed must be such that the satellite travels around the Earth once every 24 hours exactly, in other words the same speed at which the Earth rotates, and in the same direction. The time period and height can be found by equations analogous to (1.14) and (1.15), but replacing $f = mg$ with universal gravity, $f = GMm/R^2$ (Equation (1.7)). This gives the height as about 36,000 km above the Earth's surface, with a velocity of about 3.1 km s^{-1}, to maintain the balance between gravity and centripetal force. Three satellites at this height can view the whole of the Earth's surface, and be used to relay information between any two points on the Earth, making them useful for telecommunications.

Waves

Wave motion occurs in many forms in natural systems – sound waves (Chapter 6), ocean waves, electromagnetic waves (including light, X-rays, radio waves and others – Chapter 3), seismic waves within the Earth, shock waves, neural signals carried by electrical waves, description of electrons in their atomic orbits and even, hypothetically, gravity waves. Although they differ in their mechanism and medium, they all have certain basic principles in common. The essential characteristic of a wave is a periodic disturbance that can propagate energy and information without propagating material.

Box 1.5

Remote sensing orbits

While satellites can travel in any general orbit, subject to the physical constraints outlined above, for a remote sensing platform certain specific orbits have very useful properties. Low earth orbits (LEOs) are commonly used as the spatial resolution will be best at the lowest possible height, and circular orbits are preferred as the scale or images will then be similar anywhere on Earth.

For an LEO to view the entire surface of the Earth in daylight, a near polar, sun-synchronous orbit is used (see Figure 1.11). The satellite travels roughly along a line of longitude, while the Earth rotates underneath it, such that the satellite passes the Equator at the same local time on each pass. Each circumnavigation, taking around 90–100 minutes (dependent on height), views a strip of the Earth's surface from pole to pole.

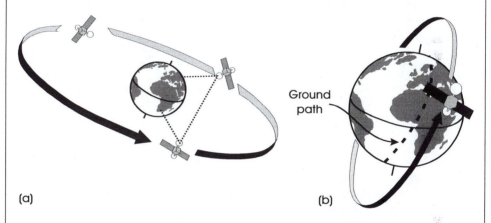

(a) (b)

Figure 1.11 *Remote sensing orbits: (a) geostationary orbit for telecommunications; (b) low earth orbit for remote sensing.*

The strips then lie together such that the whole Earth surface is covered over a number of days. This technique has the advantage that each strip is viewed at the same local time and so has the same Earth–Sun angle of illumination, allowing more consistent identification of features from their shadows and reflectance patterns. This type of orbit is used by many earth observation and meteorological satellites including Landsat and SPOT. For instance Landsat 4 and 5 are in orbits at 705 km height with a time period of 90 minutes, covering the Earth in 185 km width bands over a period of 16 days. The equator is passed at 9:45 a.m. local time on each pass.

Geostationary orbits are useful for meteorological observations, for instance producing regular, frequent images of cloud cover over a specific area such as those shown on TV weather forecasts. Because of their height the area viewed is large (virtually an entire hemisphere), but resolution is low. For more on remote sensing, see pages 128–33.

Wave characteristics

Any wave has a **wavelength** and a **frequency** that characterise it. If you are in a (stationary) boat on the ocean, you could measure the distance between successive wave crests – their wavelength, measured in metres, given the symbol λ. You could then count the number of crests passing in a minute and divide by 60 to get the frequency in Hz meaning waves per second, symbol f. Now say your boat is 12 m long, and you are watching waves with a wavelength of 3 m and frequency of 0.5 Hz (1 wave in 2 seconds). Along the length of your boat you will thus see four waves, which will take 8 s to pass the full length. The speed of the waves is thus given by 12 m divided by 8 s or 1.5 m s^{-1}.

Mathematically, we have established that the number of waves is given by $n = L/\lambda$, and the time taken to be $t = n/f$, where L is the length of the boat, giving the speed to be the distance L divided by the time. It is evident that it doesn't matter how long your boat is – the speed of the waves will always be given by the same formula:

$$
\begin{aligned}
v &= L/(n/f) \\
&= Lf/(L/\lambda) \\
v &= f\lambda
\end{aligned}
\tag{1.16}
$$

This formula is true for all waves. For light and other electromagnetic waves the speed is the speed of light and for sound it is the speed of sound; for other wave types it can vary according to the intensity of the wave. Speed also varies according to the medium travelled in – waves on treacle would move more slowly than those on water; similarly sound travels faster in water than in air, and light travels fastest in empty space.

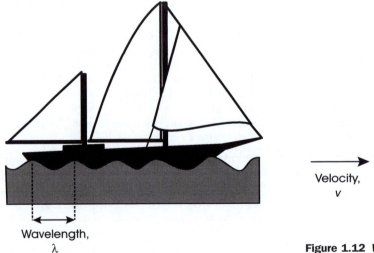

Velocity, v

Wavelength, λ

Figure 1.12 Wave characteristics.

The **period** of a wave is the time taken for a complete wavelength to pass. In the example on the boat, the period is 2 s. After a time of 1 s, there will still be four waves visible but they will have moved along the boat by half a wavelength. The crests will coincide with the position of the troughs of the original waves, and vice versa. The two waves are said to be out of **phase**, the phase being the proportion of the period of the wave that has elapsed.

There are two main wave types – **transverse** waves, where the wave travels in a direction perpendicular to the vibration, and **longitudinal** waves, where the vibration travels in the same direction as that travelled by the wave. Ocean waves and electromagnetic waves are transverse; sound waves are longitudinal while seismic waves can be both. For seismic waves these are referred to as S-waves or shear (transverse) waves and P-waves or compressional (longitudinal) waves, illustrated in Figure 1.13.

Waves carry energy, and can do so very efficiently. Remarkably, the energy lost by an ocean wave travelling across the Atlantic ocean is less than the energy that

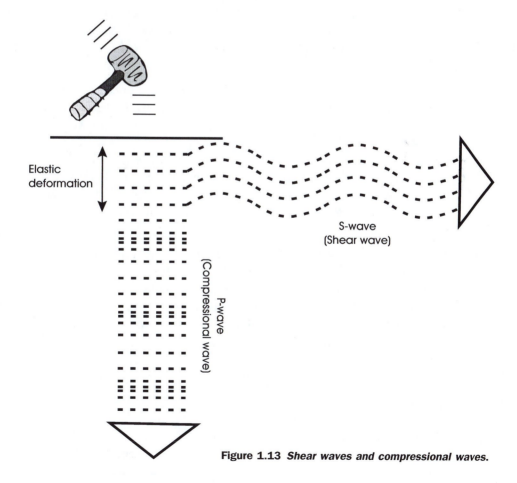

Figure 1.13 *Shear waves and compressional waves.*

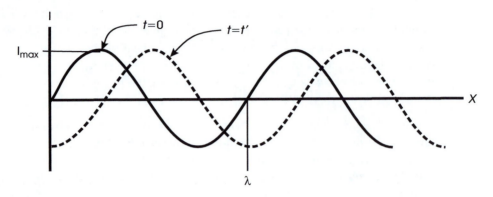

Figure 1.14 *A sine wave.*

would be lost if the equivalent amount of power were sent as electricity through an undersea cable. Light travels from the Sun as electromagnetic waves with virtually no energy loss until it meets our atmosphere. The amount of energy carried by a wave can be determined by its physical characteristics. In the case of an ocean wave this will include its height as well as frequency and speed. For light, as there is no direct analogy with height, it depends only on frequency.

The vast majority of natural waves have a shape that can be described by a **sine wave**. A typical progressive wave can be expressed by a function of the form:

$$I = I_{max} \sin 2\pi/\lambda(vt - x) \tag{1.17}$$

where the wave has intensity I, wavelength λ, speed v, with time given by t and distance by x, as shown in Figure 1.14. Note that here the sine function assumes its argument to be in radians, where 2π radians equal $360°$. From this basic equation, many other wave properties and characteristics can be deduced.

So far this discussion has considered progressive waves – that is waves that move. Waves may also be stationary or **standing waves**, such as those produced in a guitar string. These are actually formed from two waves travelling up and down the string in phase with one another, so that the resulting overall movement appears stationary. Standing waves can also appear in rivers where the current flows over a shallow uneven bed or when tidal flows and river flow interact, and on the sea where wind blows against tide.

Wave properties

Figures 1.15 and 1.16 illustrate these properties, common to many different wave types.

Transmission and absorption. Different materials will transmit or absorb waves of different frequencies. For instance the atmosphere lets through some wavelengths but not others, and X-rays pass through flesh but not bone.

Reflection occurs when a wave bounces off a hard surface. This could be an ocean wave bouncing from a sea wall, an echo from a cliff or the reflection in a mirror. If a wave is not transmitted or absorbed, its energy has to go somewhere so it will be reflected. The angle of reflection is equal to the angle of incidence, like a ball bouncing off the cushion when playing pool or snooker, marked θ in Figure 1.15. In many instances a wave will be partly reflected and partly absorbed or transmitted when it strikes a new medium, as observed in reflections from water or glass. Reflectance of a surface measures the proportion of the incident flux that is reflected.

Refraction occurs when a wave is deflected when it enters a medium at an oblique angle, because its speed is different. This explains why a spoon in a glass of water looks bent. The light travels more slowly in water than in air which causes the deflection of the ray. As well as light, sound and seismic waves commonly exhibit this property. The angle of refraction is determined by the refractive index of the medium, higher for water than air. Refractive index depends on factors such as temperature, pressure and density, so refraction will occur in the atmosphere between layers with different properties and refractive indices, or in the ground between different rock types. Snell's law states that:

$$\frac{\sin \theta}{\sin \theta'} = \frac{n'}{n}$$

(1.18)

where θ and θ' are the angles of incidence and refraction; n and n' are the refractive indices of the two media. The refractive index in this equation may be substituted by the velocities, e.g. for sound, as this is equivalent.

Diffraction is when waves bend around corners – like when ocean waves sweep around the end of a pier, and end up travelling in a different direction from those on the open sea. Sound, light and all other waves do this too under the appropriate conditions. In open conditions waves will travel in a straight line, but round a corner or through a hole they will bend and spread out. The amount they spread out depends on wavelengths, relative to the size of the hole or the sharpness of the corner. Longer wavelengths (i.e. lower frequencies) will spread out more than shorter wavelengths (higher frequencies).

Scattering occurs when a wave hits some obstacle and is broken up and redirected, as when waves pass a rock in the sea. It is a combination of diffraction around the obstacle and reflection from it. Scattered light is diffuse, as rays travel in random directions, and some will be back-scattered or returned towards its source. Scattering is made use of in measurements of water turbidity and opacity of smoke, in photometers that illuminate the target and measure how much light is scattered at right angles.

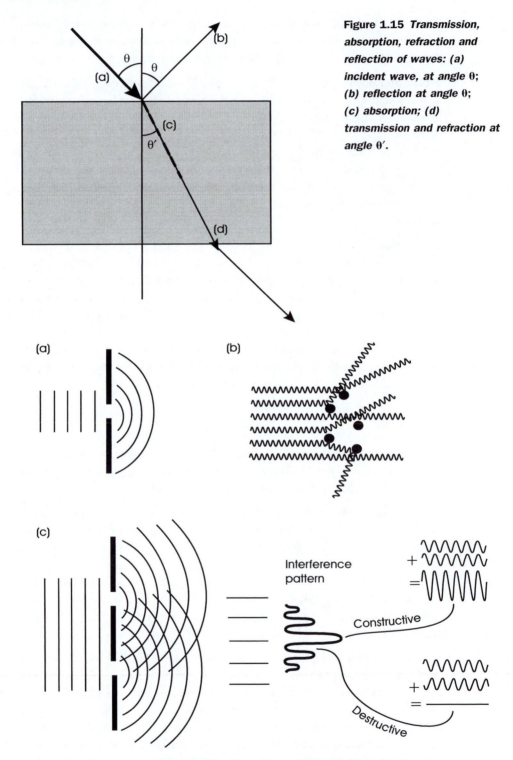

Figure 1.15 *Transmission, absorption, refraction and reflection of waves: (a) incident wave, at angle θ; (b) reflection at angle θ; (c) absorption; (d) transmission and refraction at angle θ′.*

Figure 1.16 *Wave properties: (a) diffraction; (b) scattering; (c) interference.*

Interference is a description of the patterns that occur when waves are superimposed. If you watch waves on the sea bouncing off a curved harbour wall, you may see that where the reflected waves cross incoming waves regular patterns of peaks and troughs result. This is a simple example of interference. Where the incoming and outgoing waves reinforce each other you get a higher peak; where they cancel each other out amplitude is reduced. Light and other radiation display interference patterns in a similar way. These patterns can be used to deduce the wavelength, or the differences in wavelength between two sets of waves, and is important in spectrometry.

Polarisation is to do with wave orientation, found in light and also seismic waves. In a light wave, electric and magnetic fields oscillate at right angles to one another in 3-D. In polarised light, all the fields are parallel, either horizontal or vertical. Ordinary light will have a mixture of all orientations. Polarised light may be separated by use of a polarising filter such as that used in sunglasses. Looking through two polarised sunglass lenses at right angles the view will be almost totally black, whereas at the same angle there will be little change from one lens. When light reflects at an angle from a flat surface such as water, it becomes partially polarised as the horizontally polarised light is more likely to be reflected.

Seismic waves

Seismic waves are shock waves, usually from earthquakes, causing vibrations that travel through the body of the Earth and around its surface. The body waves are of two types as shown in Figure 1.13: P-waves travel faster and can travel in solids or liquids; while S-waves move more slowly than P-waves and cannot be transmitted in liquids as they diffuse rapidly. Surface waves are transverse waves, consisting of rippling movements of the surface of the Earth. They are of two types, which are differently polarised – **Rayleigh waves** consist of vertical movements, similar to waves on the sea, while in **Love waves** motion is horizontal. Surface waves travel more slowly than body waves, but cause most of the damage associated with earthquakes.

Because of the difference in their properties, seismic body waves can be used to accurately locate an earthquake from a long distance away. To locate an earthquake, seismic monitors at at least three sites around the earthquake epicentre are used. Because of the difference in speed of the P- and S-waves, the distance from the epicentre can be found from the difference in arrival times of the two sorts of wave. The exact position can then be pinpointed by triangulation.

Seismic waves may also be used to study the interior structure of the Earth. Their velocities depend on the density and temperature of material through which they are being propagated. S-waves cannot travel through the liquid part of the core,

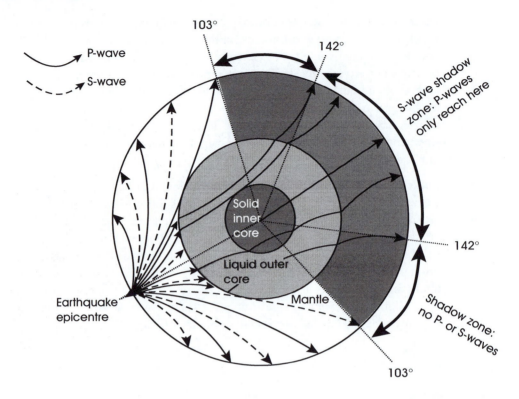

Figure 1.17 *Seismological evidence for the Earth's structure – the shadow zone.*

resulting in a shadow zone where there are no S-waves opposite the epicentre on the Earth. In addition P-waves are refracted by the liquid core – like light being refracted in water. This leads to a zone characterised by an absence of both P- and S-waves, which is found to be between 103° and 142° of latitude/longitude away from the epicentre. By this means the size and structure of the core and mantle can be inferred – details that are hard to find any other way, as you cannot just dig a hole and look.

Resonance

If you tap a wineglass with a spoon, it rings with a certain note. This note is its resonant, or natural, frequency. What you have done is set up a pattern of standing waves in the glass, with wavelengths that fit its shape. If you sing the same note loudly enough, the glass will start to vibrate, and to ring. Allegedly, opera singers can sing so strongly and purely that they can break glasses by making them vibrate in this way.

The principle of resonance is simply that if you vibrate something at the frequency at which it naturally vibrates, it will absorb most energy and vibrate most strongly, compared to a different frequency. It is very like pushing a child on a swing – gentle pushes at just the right moment at the end of the swing build up the motion higher and higher. Pushing out of phase with the swing's motion is far less effective.

Many objects have resonant frequencies, from bridges to atoms. If such an object experiences vibrations at the appropriate frequency, it will absorb energy readily. This is the cause of absorption of radiation in the atmosphere by gases – the frequency of the electromagnetic radiation is the same as a resonant frequency in the gas molecules. Resonance in larger objects can cause them to vibrate, which may result in structural damage in buildings or bridges.

During a pop concert in Finsbury Park, north London, the band Madness sang their well known crowd-pleaser 'One Step Beyond', accompanied by thousands of their audience stamping and dancing in tune with the music. Unfortunately for local residents the resonant frequency of neighbouring tower blocks was at the same frequency as the beat of the song, and the vibrations caused by all the stamping, travelling through the Earth, were sufficient to cause the entire blocks to move noticeably in time with the music. The band will probably not be asked back by that local council.

Electromagnetism

Electricity and magnetism are phenomena that we do not usually perceive directly (unless struck by lightning), yet they are important in many environmental systems and can be used in many environmental investigative techniques. Electricity provides an important form of energy, clean to use but with many environmental issues surrounding its generation and transmission. Electrical signals form the basis of our nervous systems, and some creatures use both electricity and magnetism in other ways. The Earth's magnetic field has altered over a geological timescale, leaving fields in rocks and soils that can be measured to investigate subsurface features.

Electric charge and current

Electric charge, measured in coulombs (C), is a fundamental property of some elementary particles. Electrons have a negative charge, while protons in the atomic nucleus have a positive charge. As the charge on an electron is 1.6×10^{-19} C, one coulomb is the charge on 6.25×10^{18} electrons.

Electric current is a measure of how much electricity (i.e. electrons) is moving along a conductor. For a current to flow, a circuit must be complete, so that

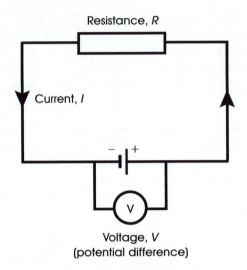

Resistance, *R*

Current, *I*

Voltage, *V*
(potential difference)

Figure 1.18 *A simple electrical circuit.*

electrons flow round in circles. Voltage, or potential difference, is how hard it is being pushed (by a battery or other power source), or more strictly the difference in electrical potential energy between two points in the circuit. The amount flowing depends on the voltage and on the resistance of the circuit. If a wire is compared to a stream, the current is how much water is flowing; the voltage is how steeply downhill the stream bed goes, and the resistance is the width of the stream bed, how rocky it is etc. The amount flowing thus depends on all these factors.

Voltage is measured in volts (V) after Voltaire, and current is measured in amps (A), after Ampère. One coulomb is defined as the charge carried by a current of one amp in one second. Electrical resistance is measured in ohms (Ω). In a simple electric circuit, the current flowing is denoted *I*, the voltage *V* and the resistance *R*. They are related by Ohm's Law:

$$V = IR \tag{1.19}$$

Mains electricity uses alternating current (AC) rather than direct current (DC). In AC with frequency of 50 Hz, both current and voltage reverse direction 50 times a second, varying cyclically between their maximum positive and negative values.

A voltage can exist without a current flowing, where a circuit is not present. Static electricity is collected electric charge where it is not moving, which can reach high voltages. Static electricity can normally only accumulate on insulating materials. On an uninsulated conductor it flows away immediately as a current.

Electric fields

Any electric charge or voltage will be surrounded by an electric field (or electrostatic field). The field is denoted by electric flux lines, which radiate outwards from the charge or current. Electric fields do not consist of any material object actually moving, but within the field, there will be a force on any other electric charge. Two charges with the same sign have fields that repel one another, while opposite charges attract. If an electric charge is near the ground, or any earthed conductor, this will distort the electric field as shown.

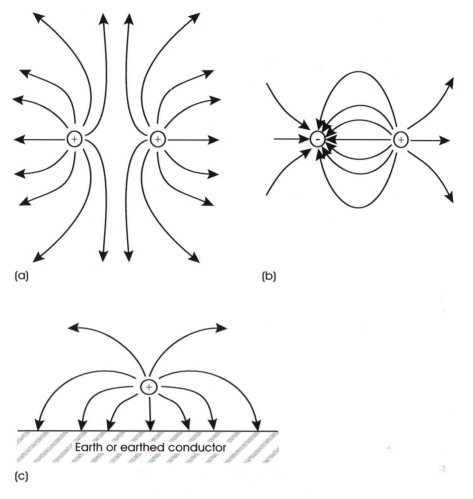

Figure 1.19 *Electric fields and forces: (a) like charges – repel; (b) unlike charges – attract; (c) effect of earth on electric field.*

In a conductor, the potential difference will be the same at any point and the electric field strength will be the same. Within the conductor, some charge will flow to one end, equalising the electric field along its length, and distorting the field. This results in one end being positively charged and one end negatively charged, giving rise to a force on the conductor. It is this effect that makes your hair stand on end during a thunderstorm; and is also important in electrostatic precipitators (Box 1.6), and in the effect of power lines on aerosols (Box 1.8).

For any conductor with a voltage applied, there will be an electric field surrounding it according to this voltage. In electric transmission lines, there is an electric field between the high voltage wires and the ground. The strength of this electric field is given by:

Box 1.6

Electrostatic precipitators

Coal- and oil-fired power stations and large combustion plant commonly use electrostatic precipitators to control the dust and particulates produced by combustion. Any non-combustible dust or ash present in the fuel passes out of the combustion chamber with the flue gases. This must be removed before being released to the atmosphere.

The flue gases pass through a chamber containing a number of flat, parallel plates hanging down, with a high voltage maintained between the plates and wire electrodes in between them. As the flue gas passes through the chamber, particulates become charged by electrons jumping from the negative electrode, and are then attracted towards the positively charged plates. Once on the plate the charge will be removed and the particles fall off into collecting hoppers below.

The process is analogous to a settling chamber (Box 1.3), with the gravitational force being substituted by the electrostatic force upon the particle. As for a settling chamber, the particle will reach a terminal velocity or drift velocity towards the collecting plate, at an equilibrium when the electrostatic force is equal to drag forces from the air. This drift velocity depends upon particle size and voltage applied, and is commonly between 0.03 to 2 m s^{-1}. High voltages combined with relatively narrow gaps between the plates can achieve efficient cleaning of particles down to below 10 μm in diameter, with up to 99.9 per cent of particulates removed. However they are expensive, requiring a considerable electrical input, accounting for their use in power plants. Dust can then be removed from the precipitator and used for purposes like breezeblocks or disposed of to landfill.

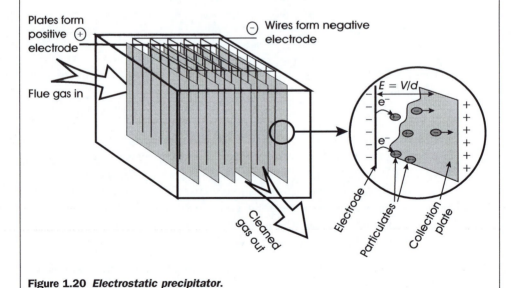

Figure 1.20 Electrostatic precipitator.

$$E = V/d \tag{1.20}$$

Electric field strength is measured in volts per metre, V m^{-1}.

Magnets and magnetic materials

Magnets commonly encountered are made from ferromagnetic materials, including iron, cobalt, nickel and alloys of these materials, which can all hold a permanent magnetic field. An alternative form of magnetism is paramagnetism, which occurs in a much wider range of materials including solids, liquids and gases. Paramagnetic materials cannot hold a magnetic field permanently, but they become magnetised when exposed to a magnetic field. If the external magnetic field ceases, they will not hold any magnetic field. The strength of magnetisation induced in a material depends on the external field strength and upon the magnetic susceptibility of the material. Magnetic susceptibility reduces with temperature, as at high temperatures, the atoms are vibrating randomly which disrupts the magnetic field.

If a magnet is heated up, at a certain temperature it will lose its magnetic field. This temperature is known as the **Curie point**, discovered by Marie Curie. Above the Curie point, the thermal energy of the atoms is too great to retain a permanent magnetic field, as the vibrations break down the field. If a piece of hot, ferromagnetic material such as iron or iron-bearing rock is allowed to cool in the presence of a magnetic field, as it passes through the Curie point it will pick up the magnetic field and become a permanent magnet. This process occurs when rocks form, and when any iron object is smelted. When a rock is in the form of hot magma, it cannot become permanently magnetised, but as it cools through the Curie point, it will pick up the Earth's magnetic field, producing **remanent** magnetisation in the rock. For magnetite, the commonest ferromagnetic mineral, the Curie point is 580°C.

Magnetic fields and forces

A magnetic field surrounds a magnet, with field flux emanating from its north pole and circling back to its south pole. As for electric fields, like poles repel one another and opposites attract – strictly they should be termed north-seeking and south-seeking, as the north pole of a magnet is attracted to the North Pole of the Earth.

Like electric fields, magnetic fields are rather tenuous entities that cannot be directly perceived by us, and do not consist of any material object or movement. Rather, the field indicates that a force would be felt by another magnet exposed to it, and there are interactions between electrical and magnetic fields – a changing magnetic field will induce an electric current (see p. 43).

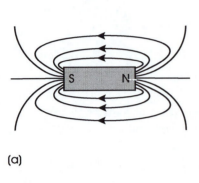
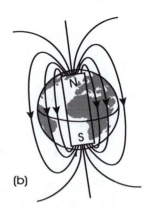

(a)   (b)

Figure 1.21 *Magnetic fields: (a) field from a bar magnet; (b) Earth's magnetic field.*

The strength of a magnetic field is measured by its flux density, measured in **Teslas (T)**. If a magnetic flux density of one Tesla is reduced to zero in one second, it will induce a voltage of one volt in a single turn of wire. This defines the Tesla, and is quite a powerful magnetic field, so often micro-teslas are used. The maximum flux density possible in iron at normal temperatures is around 2 T.

The Earth's magnetic field

The Earth has a magnetic field – as if there were a huge bar magnet inside, with north and south poles at the Earth's magnetic poles. The Earth's magnetic field is believed to be caused by fluid and electrical currents in the liquid iron core of the Earth. The liquid outer core contains convection currents and circulates due to its angular momentum from the Earth's rotation, but may be faster than the surface as the Earth is slowing down over time. As the temperature of the Earth's core is well above the Curie point, a static magnetic field could not be maintained. However, it can be shown that a magnetic field can be created by a self-sustaining dynamo effect in a rapidly rotating conducting medium, such as liquid iron. The field induces electrical currents that have a feedback effect, producing a large, stable magnetic field. This has been demonstrated experimentally in a system of rotating liquid sodium (Gailitis *et al.* 2000).

The magnetic field curves round from south to north, although it is not exactly regular. At any point (apart from the pole itself) the field points roughly north but varies with position and time, with its actual direction given by the **declination**. It also has a certain **inclination** (or dip) relative to the Earth's surface – horizontal at the equator, vertical at the magnetic poles and increasing in between. In the UK the Earth's magnetic field is inclined at about 70° to the horizontal. Its strength ranges from 25 μT at the equator to 60 μT at the poles, varying daily and over longer timescales.

The field varies cyclically and fairly predictably, due to the influence of the solar magnetic flux – a stream of charged particles from the Sun that bombards the upper atmosphere – and to changes in fluid movements within the Earth. The magnetic poles are fairly close to, but not exactly the same as, the geographic north and south poles, and their exact position varies by a measurable amount each year. Local variations also exist due to specific magnetic rock bodies. Compass navigators must take this into account – for example, in eastern Cornwall, UK in 1999 magnetic north was estimated at 3° west of true north, decreasing by 0.5° over the following four years (Ordnance Survey 1995). Close to the poles a compass would not point north at all, the direction depending heavily on the position and making the compass very limited in its use for navigation.

At irregular intervals larger changes take place – at various times in the past, the field has reversed itself, so that a compass would point south. This has happened at apparently random intervals, with periods of between 10,000 years and 1 million years between reversals. The mechanism governing field reversals is not well understood, depending on instability in the flows in the Earth's core and on changes in the solar magnetic flux. The last time was about 750,000 years ago – we may be due another one soon, but it cannot be predicted.

These magnetic reversals are known about from studies of magnetism in ancient rocks, by looking at the remanent magnetisation in the rock created when it formed. At the mid-ocean ridges new igneous rocks are forming by sea-floor spreading, becoming magnetised as they cool. As this process has been going on for millions of years, over this time the Earth's magnetic field has repeatedly reversed. The result is that rocks on the sea-floor are magnetised in 'stripes' of reversing polarity, according to when they were formed. This gives a way of measuring the rate of movement of the plates, found to be up to 10 cm per year in some parts of the Pacific.

From a detailed knowledge of magnetic reversals combined with the known movements of the continental plates, the ancient magnetic field direction and

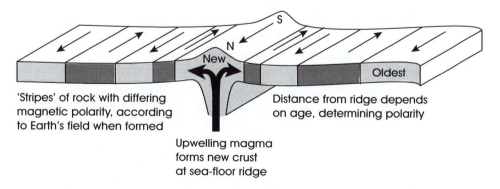

Figure 1.22 *Magnetic reversals and sea-floor spreading.*

Box 1.7

Geomagnetic surveying

Measurement of the Earth's magnetic field can provide data on subsurface features affecting the field, useful in environmental and geological investigations. In general, any rock will have a magnetic field with two components: remanent magnetism created as the rock formed and induced magnetism from the present day magnetic field of the Earth. In igneous rocks remanent magnetism is the field acquired as the rock cooled through the Curie point, while in sedimentary rocks the particles they formed from would align as they settled then become fixed. The strength of both remanent and induced magnetisation will depend upon the rock type, as those rocks containing large amounts of iron-bearing magnetite have higher magnetic susceptibility. The two fields will also differ in direction, as the remanent field depends upon the rock's position and the Earth's field at the time it formed.

These magnetic fields can be measured, often by aerial surveys although ground surveys are used for high resolution. From these observations, **geomagnetic anomalies** may be calculated, which are the difference in total field intensity between the observed and predicted value for the Earth's field, depending upon the underlying rock's magnetisation.

Geomagnetic anomalies can be used to investigate many subsurface features that affect magnetic susceptibility, or variations in rock type and structure that affect the remanent magnetisation. Features such as faults, caves and cavities or intrusions of a different rock will alter the magnetic fields at the surface, appearing as discontinuities in the geomagnetic anomaly. This technique can be useful in mining exploration, in hydrology, and in planning developments such as waste repositories. It is also useful in archaeological studies, as any disturbance or hot fire may alter the remanent magnetism, thus underground features such as tombs or hearths and fired objects such as pottery may be located. Buried magnetisable objects will also affect the field, in particular steel pipes or drums. For this reason geomagnetic surveys are used in mapping potentially hazardous waste disposal and landfill sites where paper records no longer exist (Sharma 1997: 96–101).

inclination can be calculated for any point on the Earth for any given date. This provides the **geomagnetic timescale**. Comparing the actual magnetic field found in rocks with these calculated fields provides a method for dating many rocks containing ferro-magnetic materials. Geomagnetic dating by this method provides an extra tool, that may be combined with other methods including radiometric dating (see pp. 251–3), the existence of fossils or the presence of overlying layers of rocks of younger ages, to give an accurate picture of the ages of many rocks.

Electricity and magnetism: induction

Electricity and magnetism are fundamentally connected. Electrical current creates magnetism, while changing or moving magnetic fields produce electric currents. This is known as **induction**. Electricity and magnetism combine to produce forces and motion, the basis of electric motors and generators (pp. 58–60). Electromagnetic radiation, such as light, X-rays, radio waves etc., consists of a self-propagating wave exchanging from electrical to magnetic fields, one inducing the other (pp. 115–16).

When a current flows in a circuit, it creates a magnetic field around it. A current in a wire is surrounded by a magnetic field, perpendicular to the direction of the current (i.e. in circles around the wire). For a coil of wire these circles join up to make a straight flux of magnetism through the middle of the coil. The polarity of the magnetic field depends on the direction of the current. An electromagnet, or solenoid, is a magnet created by a current in a coil of wire surrounding a metal core, which concentrates the magnetic field. It is possible to create very powerful magnets in this way. The magnet can be switched off, or reversed in polarity, simply by changing the current.

If a conductor carries a current at right angles to an external magnetic field, it will experience a force. The force tends to move the conductor in a direction perpendicular to both the direction of the magnetic field and the direction of the current.

Conversely, a moving magnet will induce a voltage in a circuit near it. The current induced will be perpendicular to the motion of the magnet. In the case of a magnet being moved into a coil of wire, a current will flow through the wire. There is work done in moving the magnet which is converted to electrical energy.

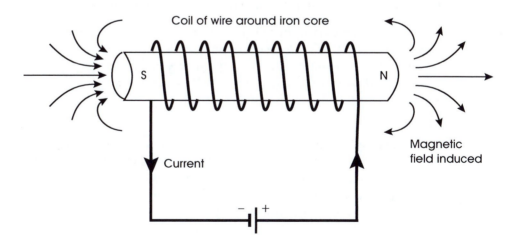

Figure 1.23 *An electromagnet.*

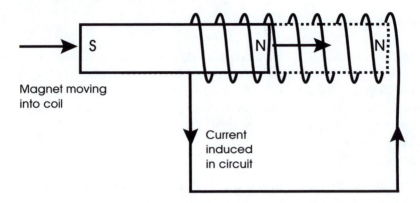

Figure 1.24 *Electromagnetic induction.*

Electromagnetism in animals and plants

There are many examples in the animal kingdom of the use of electricity and magnetism, even amongst humans – it is thought that human dowsers are subconsciously sensing magnetic fields in locating water.

Electricity is used by many small species of fish living in dark, turbid waters, such as the South American knife fish, to detect their food. The fish generate electrical discharges within their bodies and use the resultant fields to detect the conductivity of water nearby. Any change in conductivity produced, for instance by a solid object or a prey species, can be detected by receptor cells all over the fish's body. The most renowned electrical fish is the electric eel, which uses electrical discharges both for navigation through small, dark crevices, and for hunting, able to kill by electrocution. Some sharks also use electrical signals to detect prey, making them able to detect currents from movements at a great distance.

All animals use electrical signals in their nerves, to send messages around their bodies. Some plants also do this. For instance *Mimosa pudica*, the sensitive plant, produces an electrical signal when touched, sending a message along sap channels to specialised cells at the base of its leaves, triggering the release of water. This makes the leaf fold up tightly within a matter of seconds, protecting it from insects or grazers.

Magnetism can also be detected by some species. It is thought that some migratory birds and homing pigeons use the Earth's magnetic field as one of several navigation aids. This allows them to fly at night, when stars are not visible due to clouds and over oceans where there are no landmarks, without getting lost. In deep sea sediments, specialised bacteria are found that are magnetically sensitive, helping them to thrive in a dark and featureless environment.

Box 1.8

Transmission lines and human health

Electric and magnetic fields have been suggested as being harmful to humans, with concerns being voiced over a hypothesised link with childhood leukaemia in children living near to overhead transmission lines. Anecdotal evidence also exists of increased headaches, fatigue, nosebleeds, and more serious complaints including immune system disorders and cancers.

Electrical fields from high voltage transmission lines are high compared with other sources (such as household equipment), and change rapidly with the alternating current. It has been suggested that these could disrupt the electrical activity in the body in some way. However no evidence has been found to prove this conclusively, despite a number of studies. Electrical fields do not penetrate the body significantly – as our bodies conduct electricity, an external electric field would be conducted away harmlessly in our skin as with any other conductor. Magnetic fields are still an issue, with rapidly varying fields above 0.2 μT considered potentially damaging although again the evidence does not show any significant link (Day *et al.* 1999). Magnetic fields depend on the current, and thus are less if transmission voltages are very high, with household equipment contributing higher fields.

One possible mechanism is via the field's effect on airborne pollution in the form of aerosols (Fews *et al.* 1999a, b). High electric fields polarise water droplets in the air, which will then be attracted by the cables, resulting in an elevated level of aerosol droplets around the cables. Meanwhile pollutant molecules such as sulphur dioxide, nitrogen oxides, and organics can become ionised by striking the power lines. These ions then dissolve in the water droplets, and may then be inhaled by people living or working near the power line. It is believed that pollutants in this charged form may be more readily absorbed, through the lungs, than those in a gaseous or non-ionic form. The droplets may also contain bacteria, viruses, particulates, and radioactive pollutants such as radon. Hence it is possible that the health effects observed may be caused by air pollution, indirectly increased by the existence of the power line, however research is ongoing in this area.

Summary

- Newton's laws include the conservation of momentum, and the definitions of a force as rate of change of momentum and the concept of a force of reaction.

- Friction is a force that is important in many real-life situations involving motion, resisting movement and dissipating energy.

- Gravity is an example of an inverse-square law, acting between any two bodies.

- Terminal velocity occurs when the forces of air resistance and gravity are in equilibrium, applicable to many phenomena from rain to sedimentation.

- Rotational motion is described by similar equations to linear motion, applicable to climate where the Coriolis force is particularly relevant, and to orbits.

- Wave properties include transmission/absorption, reflection, refraction, diffraction, scattering, interference and resonance, which are common to many wave types.

- Electricity and magnetism produce fields that give rise to forces, and which when moving can induce one another.

- Magnetic fields form permanently in ferromagnetic materials below the Curie point, and temporarily in paramagnetic materials, processes which together with the Earth's magnetic field are used in geomagnetic surveying.

Questions

1 A cyclist accelerates at 0.4 m s^{-1} for 10 s, starting from a standstill. What velocity does it reach? How far has the cyclist travelled in the 10 s?

2 What is the momentum of a 38 tonne lorry, driving down the motorway at 60 mph? (60 mph = 27 m s^{-1}; 1 tonne = 1000 kg). If the lorry above can decelerate at 2.5 m s^{-2}, how long will it take to come to a stop? What forces are acting between the tyres and the road, and what are the sizes of these forces?

3 A monkey is hanging from a branch when she sees a hunter with a gun aiming straight at her. Just as the hunter pulls the trigger, the monkey lets go of the branch hoping to avoid the bullet. Does it work? (Hint: think about gravity acting on both monkey and bullet.)

4 A ball is thrown in the air with velocity 15 m s^{-1}. How high does it go?
(Hint: at maximum height, velocity is zero.)

5 If I drop an object on the floor, what happens to its momentum – is it conserved?

6 If you swirl a bottle of water upside down, without a lid on, what shape will the water form in the bottle as it comes out? Why does it continue to rotate and not just pour straight out under gravity? The water in the centre will move more rapidly than you swirled it – why is this?

7 A microwave oven uses microwaves with a wavelength of 4 cm. Given that their speed is the speed of light, 3×10^8 m s^{-1}, what is their frequency?

8 What is the electric field strength under a 132 kV pylon line with a height of 15 m, which runs due east–west? What would happen if a compass was carried under the lines? If no current were flowing, would this alter the electric field strength, or the effect on the compass?

Answers to numerical parts

1 4 m s^{-1}; 20 m.

2 1,026,000 or 1.026×10^6 k gm s^{-1}; 10.8 s; 380,000 N and 95,000 N.

4 11.25 m

7 7.5×10^9 Hz or 7.5 GHz

8 8.8 kV m^{-1}

Further reading

There are many general physics texts, aimed either at school physics students or non-physics undergraduates, including the following:

Physics (5th edn). J. D. Cutnell and K. W. Johnson. John Wiley, New York, 2000. Excellent applied physics text aimed at undergraduates.

Advanced Level Physics (7th edn). M. Nelkon and P. Parker. Heinemann Educational Books, London, 1994; or *Fundamentals of Physics*. D. Halliday, R. Resnick and J. Walker. John Wiley, New York, 2000. Traditional school physics texts, comprehensive and useful references.

Physics for Life. P. Warren. John Murray, London, 1988; or *Advanced Physics*. T. Duncan. John Murray, London, 2000. These are simpler and both have a lively and informative approach.

Environmental engineering textbooks are also of use, including *Introduction to Environmental Engineering*. M. L. Davis. McGraw-Hill, New York, 2000; *Foundation Science for Engineers*. K. Watson. Macmillan, London, 1993; or *Environmental Engineering* (2nd edn). P. Aarne Vesilind, J. J. Peirce and R. F. Weiner. Butterworths, Massachusetts, 1988.

Environmental Geology: Geology and the Human Environment. M. R. Bennet and P. Doyle. Wiley, New York, 1997. Good for landslides and other geological phenomena.

The Penguin Dictionary of Physics. Valerie H. Pitt (ed.). Penguin Books Ltd, Harmondsworth, 1977. A handy reference, to remind you of definitions covering a wide range of concepts.

2 Energy

Energy in its many forms is fundamental to biological life. Energy from solar radiation fuels photosynthesis, providing the basis of the food chain. Solar energy also drives the hydrological and atmospheric processes that control climate and the natural environment. The following key concepts are covered in this chapter:

- **Energy is a fundamental conserved quantity – it can never be created or destroyed**
- **Entropy governs the tendency for energy to be continually converted into lower grade forms, and keeps efficiencies less than 100 per cent**
- **Solar energy can be harnessed via many natural forms including wind, water and wave power**
- **A growing proportion of energy consumption is used for transport, with efficiencies determined by the physical processes inherent in the various transport modes**
- **Energy in the biosphere virtually all stems from photosynthesis in green plants, which also can provide a fossil fuel substitute in the form of biomass**

Human civilisation now depends on utilising large amounts of energy, from fossil fuels, renewable sources and nuclear power, in industry, transport and our everyday lives. This energy use contributes to many of the most severe environmental problems we face – climate change, acid precipitation, oil spills and radioactivity to name but four.

This chapter will cover some of the fundamental characteristics of energy, including efficiencies and entropy. Mechanical energy and power are discussed, followed by a description of some of the mechanical renewable energy sources – wind, water, wave and tidal power. Energy use in transport is then covered, and finally an ecological perspective is taken, looking at energy in photosynthesis and other natural processes.

Energy, efficiency and entropy

The First Law of Thermodynamics, the law of conservation of energy, states:

> Energy can neither be created nor destroyed, only converted between different forms.

This is a fundamental law of physics – it is always true, forever and always. So when people talk about using energy they really mean converting energy to another form.

Energy is defined as the capacity to do mechanical work, such as lifting an object against gravity. Forms of energy include: heat; light and other radiation such as radio waves and X-rays; motion, sound and vibrations; chemical energy such as in petrol, food or TNT; electricity; gravity; nuclear energy; and mass. It is possible to convert energy between any of these forms.

It used to be thought that mass was conserved, until Einstein postulated that it is possible to convert mass into energy, as occurs in nuclear reactions. The large amount of energy released is described by his historic formula $E = mc^2$, where c is the speed of light. He thus showed that mass is another form of energy.

In SI units, energy is measured in Joules (J), which will be defined later. As one Joule is quite small, more commonly used are the multiples kJ, MJ and so on.

Energy efficiency

Although energy is conserved, often we are more interested in the useful energy output from a process. Most processes converting energy are not 100 per cent efficient, as some energy is inevitably converted to waste heat or sound. The energy efficiency is the ratio of useful energy output to total energy input

$$\eta = E_{out}/E_{in} \quad (2.1)$$

e.g. a light bulb converts electrical energy into light and waste heat. A tungsten filament light bulb may only be around 10 per cent efficient, $\eta = 0.1$ by conservation of energy:

$$E_{elec} = E_{light} + E_{heat}$$
$$E_{light} = \eta\, E_{elec}$$
$$E_{heat} = (1 - \eta)\, E_{elec}$$

So a conventional 60 W light bulb produces 6 W of light and 54 W of heat.

Efficiencies can be multiplied together for processes with several stages, e.g. converting coal to electricity at 35 per cent, then to light at 10 per cent has an overall efficiency of 3.5 per cent. Almost all conversion processes produce some waste heat, so eventually virtually all the energy we 'use' in everyday life – whether in cars, lights, TVs or computers – ends up as heat.

Entropy

Entropy is a measure of disorder, also related to quality of energy. A higher entropy system is characterised by uniformity of temperature and density, with little variation in energy or potentials – diffuse heat, as a low-grade energy source, has high entropy. Low entropy is when there are extremes in temperatures, or things with distinctive characteristics and different densities, energy and state – high grade energy forms such as fossil fuels, electricity and motion have low entropy.

The Universe is gradually increasing its entropy, as it spreads out and cools down. This is described by the Second Law of Thermodynamics, which states:

> In any closed system, entropy can never decrease.

A hot bath will cool down, warming a cold bathroom in the process, until air and bathwater are both lukewarm. Within this closed system – the bathroom – energy is conserved, but entropy has increased. The heat cannot be collected from the air and used to heat the bath again, without external energy input.

The phrase 'closed system' is important, as of course the bathwater could be re-heated if topped up from the central heating boiler, adding an external source of low entropy fuel. Similarly, the Earth constantly receives input in the form of solar radiation, that supports life by providing the necessary energy, keeping entropy low.

Entropy determines whether a process is reversible – if entropy remains constant it is reversible; if entropy increases, it is irreversible. In the real world entropy always increases, albeit slightly, and thus no process is truly reversible. For instance, if I drop a bone china cup, it smashes to bits. Entropy has increased, and the process is irreversible. If I drop a super-bouncy rubber ball on a hard surface, it converts energy from potential energy, to kinetic, to elastic strain in the ball when it hits the ground, then back to kinetic and back to potential as it bounces up. It has lost some energy to heat through frictional forces with the air, and as it deforms when it bounces, so it does not bounce back quite as high. Entropy increases slightly, but the process is partly reversible – it can bounce many times before it loses all its original potential energy.

High-grade, low entropy forms of energy have more potential to do useful work, because of this irreversibility. Low temperature, diffuse heat has high entropy, while electricity has low entropy. So it is easy to convert from electricity to heat with close to 100 per cent efficiency, in an electric fire for instance. But in order to convert from heat into electricity some energy is wasted as otherwise entropy would decrease, breaking the Second Law of Thermodynamics. A considerable fraction of the heat must be converted to lower temperature, higher entropy waste heat, in order to produce a small amount of low entropy electricity. The efficiency must be less than 100 per cent, and the lower the temperature of the heat source (and so the higher its entropy), the lower the efficiency.

Motion and electricity can be inter-converted with fairly low losses either way, as they both have low entropy. Motors and generators can have efficiencies of 95 per cent or more, compared with typically 35 per cent for a thermal power station, which is looked into in more detail on pp. 109–11.

Box 2.1

How energy efficiency can exceed 100 per cent without breaking the laws of thermodynamics

Thermo-Bru brewery has four main areas of energy demand: electricity for mechanical stirring of the wort, lighting, and other appliances; high temperature process heat at 300°C to convert the barley into malt; process heat at 100°C to boil the fermenting liquor, and low-temperature heat to maintain the fermenting vessels' temperature at 40°C and provide space heating and hot water for their offices.

The energy demands for the four are:

Electricity	5,000 MWh (18,000 GJ)
Heat – 400°C	25,000 GJ
Heat – 100°C	15,000 GJ
Heat – low temperature	10,000 GJ
Total demand	68,000 GJ

All power is provided on site from a gas-fired combined heat and power plant. This produces electricity with a 35 per cent efficiency, requiring 51,430 GJ of gas to provide the required electricity demand. Waste heat from the gas furnace, amounting to 33,000 GJ, is recovered at 80 per cent efficiency, providing ample high temperature heat to roast the malting barley. Once roasted, the malt is cooled via a heat exchanger, recovering 75 per cent of its heat which provides 18,750 GJ, sufficient to boil the fermenting liquor. This is again cooled via a heat exchanger at 70 per cent efficiency, providing 11,250 GJ for the low-temperature heat demands.

Total primary energy input is 51,430 GJ

Total useful energy output is 68,000 GJ

Efficiency = 68,000/51,430

= 132 per cent,

and no physical laws have been broken.

While this example may be rather far-fetched, the principle of a 'heat cascade' from the highest to lowest temperatures, minimising waste heat, means energy can be used again and again, greatly increasing efficiencies. There is such a thing as a free lunch!

Efficiency of power stations could be far greater than this if some use is made of the 'waste' heat, as in a combined heat and power (CHP) plant – see Box 2.1. Combined heat and power is common in some countries such as Scandinavia and parts of northern Europe, where power stations or large industrial plant are linked to district heating systems that provide piped hot water to apartment blocks or housing estates. CHP provides massive opportunities to reduce primary energy use and associated pollution, in many cases without increasing costs, although it is dependent upon the demand for electricity and heat matching, and being physically close as heat cannot be transported as easily as electricity.

Entropy and the environment

In a philosophical way, many of our environmental problems are characterised by large increases in entropy in the natural environment. It is because these changes are irreversible that our society is unsustainable. We extract natural resources to produce goods, and when they are worn out we throw them away and start again. We don't commonly re-use those resources, which are now dispersed through landfill sites or incineration. These are high entropy systems that would require high energy input to reclaim the materials. In the process we rely on the throughput of large amounts of non-renewable energy.

In many ways, mankind's interaction with Earth is to turn a highly complex, diverse system (low entropy) into a much simpler degraded environment (high entropy), in a way that is essentially irreversible due to the increases in entropy taking place. The natural systems are sustainable because they are reliant only on the energy from the Sun to maintain their low entropy, while we have increased the energy throughput by using fossil fuels. In the process we are creating our own very complex physical and social environment which has low entropy, but is dependent upon these energy flows, and on degradation of natural systems.

Table 2.1 *Environmental issues and entropy*

Low entropy state	Higher entropy state
Fossil energy resources in concentrated form underground	Energy degraded to heat; resources reduced to CO_2 and dispersed through the atmosphere
Minerals and metals in concentrated form underground	Minerals and metals dispersed through minewaste and final disposal of goods after consumption, energy used in processing
Land covered by highly diverse ecosystems with many species	Much land in monoculture agriculture sustained by high energy input, many species extinct
Surface waters kept pure by balance of biological and physical systems	Rivers and oceans polluted by low concentrations of many toxic and carcinogenic substances
Soils developed over a long time, containing a balanced community of animal, plant and bacterial life	Soils reduced by pesticides/herbicides to an inorganic substrate needing energy-intensive artificial fertilisers to be productive, or damaged by erosion

Overall therefore entropy is increasing much faster. Much of environmental awareness is about recognising this process, and making value judgements about whether what we are destroying is a price worth paying for the material world we are creating.

Mechanical work – forces and energy

Mechanical work is readily experienced as an energy form, whenever you exert a force to lift, push or pull something. A force only uses energy if it does some work, i.e. if it moves something such as lifting an object against gravity, stretching an elastic band or accelerating a car. Many systems have forces in equilibrium, where no work is being done and so no energy is expended.

For instance if you stand holding a book at arm's length but not moving it, you are not doing any work and so theoretically not using energy. If you don't believe this, imagine the book could be sitting on a bookshelf instead – which would not need any energy, as bookshelves don't need to be plugged into the mains. In reality you are expending some energy as your muscles are continually expanding and contracting to maintain the force. But you only do mechanical work if you lift up the book, moving it against the force of gravity, when the energy is the work done by a force moving through a certain distance:

$$\text{Energy} = \text{force} \times \text{distance} \tag{2.2}$$

The Joule is defined by this relationship. One Joule is the work done by a force of 1 N moving through 1 m, equivalent to lifting a mass of approximately 100 g (about the size of an apple) up by 1 m.

$$1\ J = 1\ N \times 1\ m$$

Energy, power and units

Power is a measure of energy per unit time. Power is instantaneous, while energy is used over a period of time. For example, a car may have higher power in a lower gear, and so may accelerate faster in third than fourth. So after driving for ten minutes in third gear, the car would probably have used more fuel – more energy – than if it were in fourth gear.

Energy and power are therefore related by:

$$E = Pt \tag{2.3}$$

where E = energy, P = power, t = time it is running for.

Energy and power can be distinguished by their units: power is measured in Watts (W), or kW and other multiples. 1 W is equivalent to 1 J s^{-1}, so 1 J is the energy produced from a power of 1 W running for 1 s.

Electricity is generally measured in kilowatt hours, kWh (sometimes just called a 'unit' of electricity). 1 kWh means *1 kW for 1 hour* – the energy from 1 kW power running for 1 hour, as in a single bar electric fire for an hour. The kWh is a measure of energy, while the kW is a measure of power.

The energy in 1 kWh can be calculated using Equation (2.3), $E = Pt$:

$$\begin{aligned} 1 \text{ kWh} &= 1 \text{ kW} \times 1 \text{ h} \\ &= 1,000 \text{ J s}^{-1} \times 3,600 \text{ s} \\ &= 3,600,000 \text{ J} \\ &= 3.6 \text{ MJ} \end{aligned}$$

Many other units are commonly used to measure energy and power by energy utilities, engineers or in everyday life, listed in Table 2.2. For instance, the calorie is the energy needed to raise the temperature of 1 gram of water by 1°C.

Table 2.2 *Units of energy and power and their definitions*

Unit	Definition	Equivalent in Joules
Joule, J	Force of 1 N moving through 1 m	1 J
kWh	1 kW for 1 h	3.6 MJ
calorie, Cal	Heat required to raise temperature of 1 g water by 1°C	4.2 J
British Thermal Unit, BtU	Heat required to raise the temperature of 1 lb of water by 1°F	1,055 J
Therm	100,000 BtU	105.5 MJ
m^3 of gas	1 m^3 natural gas at a given temperature and pressure	40 MJ
Tonne oil equivalent, TOE	1 t crude oil	45 GJ
Tonne coal equivalent, TCE	1 t standard coal	30 GJ
Electron volt, eV (used in atomic physics)	Energy gained by an electron moving through electric field of 1 V	1.6×10^{-19} J
Watt	1 J per second	$1 \text{ W} = 1 \text{ J s}^{-1}$
Horsepower, HP	Measure of power; historical definition	746 W or 746 J s^{-1}

1 Cal = 4.2 J. Bear in mind that in diets, calorie is used to mean kCal, so a diet of 2,000 calories a day actually means 2,000 kCal = 8,400 kJ.

Any of these energy units can be converted into Joules by multiplying by the equivalent in Joules. For instance, if your gas bill shows you to have used 356 m^3 of gas with an energy value of 40 MJ m^{-3}, the amount of energy used is:

$$E = 356 \times 40$$
$$= 14{,}240 \text{ MJ}$$

To convert into kWh, divide by the figure for a kilowatt hour in Joules, 3.6 MJ kWh^{-1}:

$$E = 14{,}240/3.6$$
$$= 3{,}956 \text{ kWh}$$

Kinetic energy and potential energy

Kinetic energy (KE) is the energy of a body due to its motion. When an object accelerates under a force f, Equations (1.1) and (1.4) from Chapter 1 tell us:

$$f = ma$$

and

$$v^2 = u^2 + 2as$$
$$= 2as \quad (\text{as } u = 0)$$

Substituting into Equation (2.2):

$$E = fs$$
$$= mas$$
$$= \tfrac{1}{2}mv^2 \quad (\text{as } as = \tfrac{1}{2}v^2)$$

This is the kinetic energy of an object at velocity v. It doesn't matter what the acceleration was or whether it was constant or in a straight line, due to the principle of conservation of energy: any two bodies with the same mass and velocity must have the same kinetic energy, however they got there.

Potential energy (PE) is energy due to position or height, e.g. energy gained by a skier by riding up a chair lift allows them to then ski down again, converting the PE gained into KE. If a mass is raised against gravity to height h, the force on it is $f = mg$ from gravity.

$$\text{energy} = \text{force} \times \text{distance}$$
$$= mgh$$

Box 2.2

Why don't we shoot CO_2 into space and solve the greenhouse effect?

The **escape velocity** from the Earth's surface is approximately 11 km s^{-1} – this is the minimum velocity needed to escape the influence of Earth's gravity so an object won't fall back to the ground again (p. 26). To get to this velocity would require a minimum of the kinetic energy from Equation (2.4):

$$KE = \tfrac{1}{2}mv^2$$

For a mass of $m = 1$ kg of CO_2:

$$\begin{aligned} Energy &= \tfrac{1}{2}mv^2 \\ &= \tfrac{1}{2} \times 1 \times 11,000^2 \\ &= 60.5 \times 10^6 \text{ J or } 60.5 \text{ MJ} \end{aligned}$$

In fact it would take considerably more, as efficiency would be less than 100 per cent, and there would be friction in the atmosphere slowing it down. The CO_2 would also require energy to compress it, and the pressurised canisters to hold it would add to the weight.

One kg of coal produces about 3 kg CO_2 and contains 29 MJ of energy.

To shoot this into space using 60.5 MJ per kg would take 181.5 MJ, which is over six times as much as the energy in the fuel in the first place.

This is its potential energy. We have now derived two important equations:

$$KE = \tfrac{1}{2}mv^2 \tag{2.4}$$

$$PE = mgh \tag{2.5}$$

Kinetic and potential energy can be converted to and from one another with great efficiency. In an ocean wave, water molecules are moving up and down in the swell, converting KE to PE and back again, with very little energy loss.

In fact *all* energy is either kinetic or potential energy. Heat is the kinetic energy of molecules vibrating; electricity is the movement or potential of electric charge; radiation is a form of oscillating electromagnetic potential; chemical energy is due to the electrical potential between atoms; nuclear is due to the nuclear potential between neutrons and protons, and so on.

Electrical energy

There are many ways to convert energy into electricity. Electrical generators as found in power stations convert mechanical energy into electrical energy by electromagnetic induction. Electricity can also be converted from radiation, mechanical or thermal energy directly. Photovoltaics (pp. 69–72) convert light into electricity. Certain crystals will produce a voltage when they are squeezed – mechanical energy from the squeezing force gives the electrons enough energy to conduct, known as piezoelectricity. This is the principle behind piezoelectric gas lighters, common on gas cookers. Piezoelectricity has been used in earthquake detection monitoring. Lastly a thermocouple is a device that produces a current according to the temperature of two junctions between different metals. It can be used to measure temperature electrically, in many environmental monitoring applications where a conventional thermometer would not be practical.

Electric power and transmission

Electrical power is measured in Watts, where $1\ \mathrm{W} = 1\ \mathrm{J\ s^{-1}}$ (like any sort of power). The power of a current of 1 A with a voltage of 1 V is 1 W.

$$\mathrm{watts} = \mathrm{volts} \times \mathrm{amps}$$

$$P = VI \tag{2.6}$$

It takes energy for electricity to flow through a resistance, which will heat it up, and any electrical appliance has got a resistance. This may be like the element in an electric kettle, designed to have high resistance and heat up; or like the resistance of the transmission lines that carry power, designed to have as low a resistance as possible so they do not heat up. By combining Equation (2.6) with Ohm's law, Equation (1.19), power can be expressed equivalently in terms of voltage or current as

$$P = I^2R = V^2/R \tag{2.7}$$

For a given voltage therefore, the lower the resistance, the higher the power. An electric cooker would have lower resistance than a light bulb for instance, allowing more current to flow and producing more power.

In a transmission line, the wire used has some small level of resistance. Over long distances, this leads to loss of electrical power by heating up of the transmission wires. If a transmission line is providing a power level P, this is given by $P = IV$. The higher the voltage, the lower the current flowing to provide a certain power. The power losses in transmission are given by:

$$p = I^2r \tag{2.8}$$

where r is the small resistance of the wires. This increases as the square of current, and so to reduce losses the current should be as small as possible, therefore the voltage should be as high as possible. For this reason, transmission in the National Grid is at high voltages, up to 400,000 V.

To change the voltage down to the safer level of 240 V or 110 V used in the home, transformers are used. Alternating current (AC, see p. 36) is much easier to transform, which is the main reason why AC is used in power networks. There is a small power loss each time it is transformed, but this is offset by the advantage in reducing transmission losses.

Power station capacity

Power station sizes are usually expressed in megawatts, (MW), i.e. in terms of power rather than energy. The amount of electrical energy produced can thus be found from Equation (2.3), $E = Pt$, for the number of hours in the year that the station is operating.

If time t is expressed in seconds (SI units), with P in MW this would give energy in MJ. But if t is given in hours, this will give the energy in MWh (Megawatt-hours) – making it simple to convert to kWh, the commonest electrical unit, or other multiples.

The time generating can be found from the **load factor**, which is equivalent to the percentage of the time the station is on full power. For annual output, the number of hours in a year is 8,760. For instance, a coal-fired power station with a capacity of 1,200 MW operates as baseload with a load factor of 85 per cent. The energy produced in a year is:

$$E = Pt$$
$$= 1{,}200 \times 8{,}760 \times 0.85$$
$$= 8{,}935{,}200 \text{ MWh}$$
$$= 8.9 \text{ TWh or } 8{,}935 \text{ million kWh}$$

where 1 TWh is 10^{12} W for one hour, i.e. a billion kWh.

Electric motors and generators

The principle of motors and generators is similar. A coil of wire is held in a strong magnetic field (from either a permanent magnet or an electromagnet). The coil is on an axle that allows it to turn without breaking the circuit. If the coil is turned, its motion perpendicular to the magnetic field will create a current in the wire by

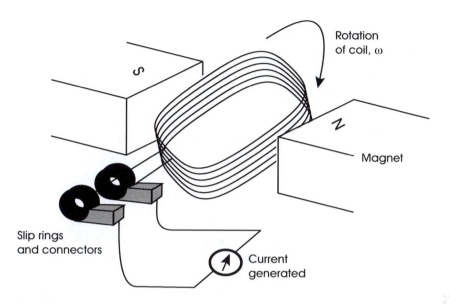

Rotation
of coil, ω

S

N

Magnet

Slip rings
and connectors

Current
generated

Figure 2.1 *Electric motor or generator.*

induction. After half a turn, the direction of rotation of the coil is reversed, and so the direction of the current is reversed, producing an alternating current (AC). This is a simple AC generator.

Using exactly the same apparatus, if a current is applied through the coil, it will create a force on the coil that is perpendicular (in 3-D) to both the current and the magnetic field. An AC current applied will thus turn the coil continuously – a simple AC motor.

A slightly more sophisticated motor or generator would have three coils superimposed, so that the movement is even over the rotation, and the problem of a 'dead spot' where there is no torque (turning force) is avoided. For a commercial electrical generator, although the principle is identical, being very large scale it is more practical to turn the magnet rather than turning the coil of wire. This obviates the need for brushes to conduct large amounts of power through moving connections, that would result in resistive losses.

Virtually all power stations in use today – whether fossil fuel, nuclear or renewables – generate power by the same principle of turning magnets in a coil of wire to induce a current. The only difference between electrical power stations is in the source of the force to turn the magnets. It may be thermal, via steam or gas turbine plant from fossil fuels, biomass, solar or geothermal (see Chapter 3) or a nuclear core (Chapter 7); or it may be mechanical, from the motion of wind, waves or water. But the electrical end of the process is identical.

The magnet being turned is an electromagnet, as it needs to be very powerful, although the amount of power it consumes is relatively small. It is surrounded by three coils of wire each at 120° to one another. Turbines in thermal power stations convert energy from high-pressure steam into rotational movement of this magnet, while in wind or water systems the magnet is turned directly. The movement induces a current in each of the coils, and the magnet is turned 50 times per second, giving rise to the 50 Hz mains frequency.

A modern electrical generator is very efficient – 99 per cent of the rotational movement is converted into electrical power.

Renewable energy

Renewable energy sources almost all stem from the Sun – with the exceptions of geothermal and tidal energy. Solar energy can be collected directly, by solar water heaters, solar collectors that concentrate heat and convert it into electricity, or photovoltaics that convert light directly into electricity. Other renewables use wind, waves, rain via hydro-electric dams, or biomass and woodfuel that all have as their source solar radiation. In fact coal, gas and oil also came from the Sun as it was the Sun's energy that grew the ancient biomass that created fossil fuel reserves.

In wind, wave and hydro-electric power, the Sun's heat causes convection currents that power weather systems, and evaporates water that falls as rain. The energy is then converted from kinetic energy of the wind and waves, or the potential energy of a head of water, via a generating turbine into electricity.

Biomass and wood convert solar energy by photosynthesis, which has a very low efficiency – less than 1 per cent of solar energy falling on a hectare of vegetation can be converted into biomass for use as fuel.

The only forms of energy we use that are not forms of solar power are geothermal power – energy from the Earth's internal heat; tidal power, coming from gravitational forces (from Earth, Moon and Sun) and nuclear power.

Renewable energy has the advantage of not producing carbon dioxide, and is one essential component in any strategy to reduce the risks of climate change. While biomass does produce CO_2 when burnt, it only releases the same amount as it took in when growing, and so is carbon-neutral over the lifetime of the tree or other plant. This contrasts with fossil fuel, which releases carbon that may have been fixed hundreds of millions of years ago when the Earth's atmosphere and climate were very different from today's.

This is not to say that renewable energy is without environmental impact, as each source has some impact which may include emission of other air or water pollutants, visual impact, damage to wildlife and ecosystems or noise, some of which are discussed below. These impacts however are generally local and short-lived, and must be compared with problems of conventional electricity generation, often international in scope and long-lasting.

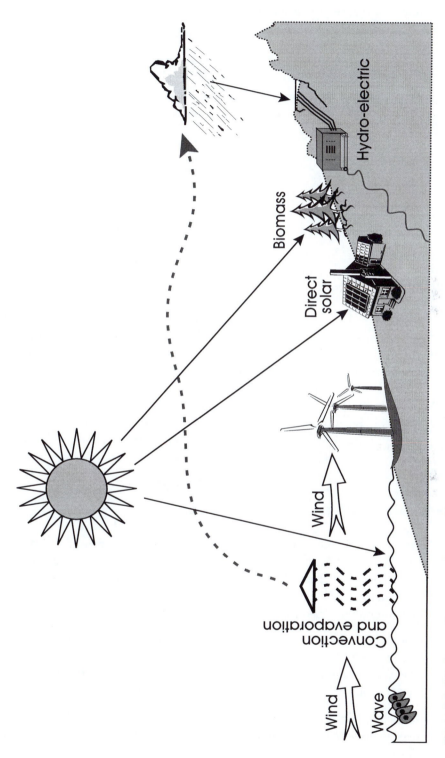

Figure 2.2 *The Sun as the source of renewable energy resources.*

Table 2.3 *Renewable energy resources*

Renewable source	Resource base TW	Recoverable TW
Solar radiation	90,000	1,000
Wind	300–1,200	10
Biomass	30	15
Wave	1–10	0.5–1
Hydro	10–30	1.5–2
Tidal	3	0.1
Geothermal	30	1
Total world energy consumption	13	

Renewable resources

As Table 2.3 shows, the amount of energy from the Sun and from the other renewable sources is very large compared with the amount of energy used by mankind. However much of this energy is either too diffuse (i.e. high entropy) or too inaccessible to use as fuel. When the recoverable resources are considered they are still very significant – a combination of these sources could easily provide all our energy. However there are other constraints, the main one being financial. Currently renewable sources account for 18 per cent of world energy consumption, mainly in the form of woodfuel for heating and cooking in developing countries, and hydro-electric power. In Europe, only 4.5 per cent comes from renewables.

Hydro-electric power and potential energy

Hydro-electric power or HEP is the production of electricity from the energy in moving water. Water mills have been used for thousands of years, but their adaptation to produce electricity was first achieved in the late nineteenth century. The energy produced comes from the potential energy of the water, which travels down through a pipe from a height, to turn a turbine.

For an HEP station with a head height h, the potential energy of the water is given by Equation (2.5), $PE = mgh$. The power as energy output per hour can then be calculated from the flow rate of the water and the station efficiency. Efficiency depends on friction in the pipes, and efficiency of conversion to electricity.

In a pumped storage system, cheap off-peak power in the middle of the night is used to pump water up a mountainside into a lake. This can then be used to

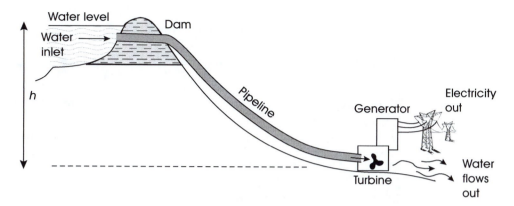

Figure 2.3 *Hydro-electric power station and pumped storage.*

produce electricity when power demand is high, during the day, and to even out demand overall. The efficiency of this process is fairly high – up to 80 per cent of the amount of off-peak electricity used in pumping can be generated when needed. However it is expensive, as the capital costs are high and the system is only generating for short periods. The electricity it produces has been generated once, then used in pumping, then generated again, so the overall cost per kWh would be at least three times the normal generating cost.

Environmentally, HEP has a poor record, as many large projects have been constructed requiring dams that flood large land areas, resulting in loss of productive land or forest and people being evicted from their homes and villages. However, small-scale plants are now being built that do not require dams or lakes, but benefit from being renewable and having no emissions. These plants provide electricity for villages and towns in many remote mountainous areas in the developing world, and supply power to the grid in Europe.

Wind power

Modern wind turbines owe more to aeronautical engineering than to the windmills used historically to grind corn. Turbine blades are aerodynamically designed, similar to an aircraft's wing. Wind farms are now becoming a familiar sight in upland areas of many countries, providing a zero-emission source of electricity that is rapidly becoming cost competitive with fossil fuels.

A wind turbine takes its energy from the KE of wind passing through its blades. For a turbine with area A swept by the blades, the amount of air passing will be given by the volume of a cylinder of wind, with surface area of its face A and length v, the distance travelled by the wind in a second.

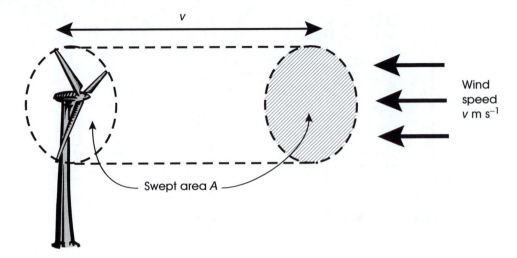

Figure 2.4 *Energy in the wind.*

$$m = \text{volume per second} \times \text{density}$$

$$= Av\rho$$

where v is the wind speed and ρ is the density of air. The KE of this mass of air from Equation (2.4) is then given by:

$$\text{KE} = \tfrac{1}{2} m v^2$$

$$= \tfrac{1}{2}(Av\rho)v^2$$

$$\text{KE} = \tfrac{1}{2} A\rho v^3 \tag{2.9}$$

This is the energy in the wind. Note that as energy depends on v cubed, doubling the wind speed will mean 2^3 or 8 times as much energy. Not all of this can be used, as the air does not stop moving entirely as it strikes the blades. Some energy is carried away by wind leaving behind the turbine. This loss determines the power coefficient – typically about 40 per cent of the energy in the wind is converted into blade movement. The energy output will then depend on the efficiency of converting this blade movement into electricity. For an actual wind turbine, the output will be less than this, as it will have a **rated capacity**, which is the maximum electrical output, and it will cut out at very high wind speeds to avoid damage.

Two important points emerge from this. First, because of the v^3 relationship, wind turbines need to be situated on very windy sites to maximise their output, as a small increase in wind speed produces a large increase in output. The same turbine on a site with average wind speeds 10 per cent higher could produce 33 per cent more power – at least when producing less than the rated capacity. This is why wind farms are generally on exposed hilltops or coasts in remote areas, not in the middle

Plate 2.1 *Offshore wind farm at Vindeby, Denmark.*
Source: photo by courtesy of Bonus Energy A/S, Brande, Denmark

of cities – and they are very often bound to be visible for long distances because of this, which has added to the conflicts over their development. Offshore wind farms such as that at Vindeby, Plate 2.1, can benefit from high wind speeds without imposing such a visual intrusion.

Second, the average output of the turbine will be considerably less than its rated capacity, because of the variability of the wind. The rated capacity cannot be directly compared with the capacity of a coal-fired power station for instance, which could be running almost all the time at full capacity. For this reason, wind farm capacities (and other intermittent renewable sources) are often expressed as **declared net capacity** (DNC), which is defined as the equivalent capacity of thermal generation that would produce the same electrical output if operating 100 per cent of the time, and is around 40 per cent of the rated capacity.

Tides and tidal power

A tidal power plant consists essentially of a barrage enclosing an estuary, such that the estuary may be filled at high tide, then the water released through generating

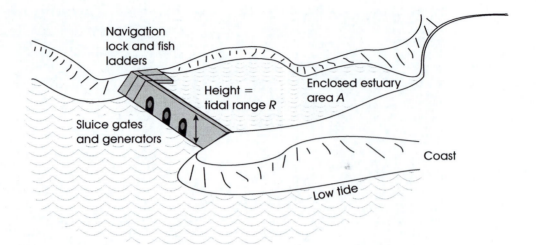

Figure 2.5 *Tidal power energy.*

turbines (very similar to those used in hydro-electric power plants) at low tide. In some designs power can also be generated on the incoming (flood) tide, as the estuary fills. The source of energy in tidal power is the gravitational pull between Earth, Moon and Sun. The tiny amount of energy lost in friction on the sea bed as the tides move to and fro (and through tidal generators) is causing the Earth to slow down slightly. As the kinetic energy of the Earth is so vast however, this is not generally considered a major environmental impact of tidal power schemes.

The amount of energy available to generate power depends on the tidal range and the area of water behind the barrage. The maximum amount of energy available to generate power is given by the difference in the potential energy of the water trapped behind the barrage at high and low tide. This potential energy is given by Equation (2.5), $PE = mgh$ where m is the mass of the water, g is gravity and h is the height difference. For a tidal estuary of height R (the tidal range) and area of water trapped A, the mass of water will be its volume (AR) times the density of water (ρ), and the average height fallen will be $R/2$:

$$m = \rho AR$$

$$h = R/2$$

$$PE = mgh$$

$$= (\rho AR)g(R/2)$$

$$PE = \tfrac{1}{2}\rho AR^2 g \tag{2.10}$$

This is the maximum theoretical energy available per tidal cycle. In fact, the amount of power generated will be considerably less than this, as to use all this potential would mean emptying the entire estuary instantaneously at low tide.

To smooth power production, the estuary would normally be emptied gradually over a period of 3–4 hours on each tidal cycle, which means that a lot of power generation occurs when the difference in height is less than the full tidal range, reducing overall output.

Tidal range is a critical factor in determining whether an estuary would be useful for tidal power generation, because its dependence on R^2 means a small increase in tidal range gives a larger increase in potential energy available. The gravitational forces that cause tides result in a relatively small tidal range in the open oceans, of around 0.5 m. At coasts and in estuaries the range can be much greater. As the tidal swell approaches land, the water gets shallower and the moving water banks up, increasing its height. In addition to this amplification, resonance plays an important part. If an enclosed body of water has a natural frequency that is equivalent to a whole number of tidal cycles, a standing wave can be set up, bouncing to and fro across the ocean like water sloshing in the bath. Sloshing at the bath's natural frequency can create large waves, and likewise where oceans happen to be the right size, the open water tidal cycle gets amplified greatly. This natural frequency depends on the width and depth and the shape of the enclosed water. The Atlantic ocean between the USA and Europe happens to be just the right width for this resonance to occur, amplifying the natural open water tidal range from 0.5 m to typically 3 m at the coasts. Other enclosed bodies of water including the North Sea, the Bay of Fundy in Canada and the Irish Sea/Bristol Channel/Severn Estuary complex set up further local resonances, amplifying the range further. This results in a tidal range at Bristol of over 11 m, making it one of the highest in the world. At other locations such as parts of the Pacific where resonance does not occur, tides can be almost unnoticeable, or in some cases only occur once a day.

The ideal estuary would not only need a high tidal range, it would have a large enclosed area with a narrow neck across which to build a barrage. These local factors make the economics of tidal power very site specific.

Tidal power may not produce any carbon dioxide or other air pollutants, but it is still one of the more environmentally damaging of renewable energy forms. The barrage itself would be extremely visually intrusive and present a potential hazard to migrating fish, especially on a large scheme such as the Severn. More crucially, sensitive ecosystems around mudflats that support large numbers of wading birds rely on the existing tides and siltation patterns which would be totally disrupted by the changes in water flow regime. Minimum water levels, waves, turbidity and salinity would also be affected, and while some of these effects could in theory increase biodiversity, existing ecosystems would be put under huge stress.

Energy in waves and wave power

Ocean waves are created by the wind on the surface of the water, building up height and energy over long distances. The size of an ocean wave depends on the wind speed, how long it has been blowing for, and the distance of water over which it has been blowing over which the waves build up, known as the **fetch**. Very large waves are often encountered in the southern oceans where winds can, and do, blow all the way around the Earth without meeting land, building up waves to their maximum height for the wind speed – individual waves of 30 or 40 m are not unknown under the right conditions.

Deep water waves are transverse waves, with individual water molecules moving in vertical circles but overall not moving forwards. They have a regular sine-wave shape, which can be described simply in terms of velocity, wavelength (or time period), and the acceleration due to gravity g. Under normal conditions wave patterns are more complicated with many waves of different sizes and directions superimposed on each other at once. As the water becomes shallower (less than 50 m for large ocean waves), near land or on a reef or sandbank, the wave starts to 'feel the bottom' and its shape and speed alter. For depths of less than a quarter of the wavelength, the speed no longer depends on the wavelength, but is governed by the water depth. Water molecules slow down with friction against the sea floor, losing energy, and then when the wave breaks the water moves forward and up the beach, dissipating its remaining energy.

The energy in a wave is being constantly interchanged between potential and kinetic energy forms, with potential depending on height and kinetic on velocity. There is most energy in deep water waves, as energy starts to be dissipated as soon as the wave feels the bottom. The energy of a deep water wave expressed as power in watts per metre of wavefront can be shown to be given by:

$$p = \frac{\rho g^2 H^2 T}{32\pi} \tag{2.11}$$

where ρ is the density of water, which for sea water is 1,025 kg m^{-3} and H is the height of the wave. To a reasonable approximation this simplifies to:

$$P \approx H^2 T \tag{2.12}$$

where P is the power in kW per metre of wavefront, H is the height in metres and T is the time period in seconds. So for a typical Atlantic swell of 3 m with a 9 s time period, this gives 81 kW m^{-1}, which makes it an attractive energy resource, compared with, say, the average solar input in these latitudes of about 100 W m^{-2}. Closer to shore this would drop as the energy starts to dissipate, reducing height and/or time period. In a typical winter storm with waves averaging 11 m and a 13 s time period, it would increase to around 1,600 kW m^{-1}. It is evident that any wave energy device must be built to withstand these extreme energy densities without damage, but also that most of the energy in a storm will not be convertible into electricity as a device designed to generate 50–100 kW could not possibly adapt to produce 1,600 kW.

There are many devices designed to extract some of this energy for electricity generation, which can be divided into offshore and onshore. The offshore devices include a variety of tethered, shaped buoys and rafts that move with the wave relative to either their mooring or to some other part of the system, and generate energy from the movement of hydraulic pistons. Onshore devices include oscillating water columns and tapered channel (TAPCHAN) systems that use the movement of the wave amplified through a channel or column either to move a hydraulic piston or to force air or water through a turbine.

Offshore devices have the advantage of tapping into the largest resource, being only limited by the depth of water and the cost of electrical connections to the shore. They are also one of the most environmentally benign forms of electricity generation, being virtually invisible from the shore and harmless to marine bird and fish life, although they may present a minor hazard to shipping. They would have little impact on the erosion or siltation of shorelines that can be important to marine ecosystems, as they would not significantly reduce the power of storm waves. The main disadvantage of offshore devices is cost – they have to exist in an extremely hostile environment, battered by the massive forces of storm winds and waves, corroded by salt and very inaccessible to maintenance crews.

There is currently more interest in onshore devices, which are rather more accessible although still subject to winter storms. This makes construction and maintenance simpler than offshore and decreases costs, but the resource is more limited and site specific. They also have significant environmental impacts, as a large scale onshore wave energy converter would involve large concrete structures along lengthy stretches of coastline, probably in remote and beautiful areas such as western Scotland or Norway.

Photovoltaics

Photovoltaics (PVs), also known as solar cells or photoelectric cells, work by the photoelectric effect, which was explained by Einstein. PVs are made from semiconductors, materials that lie between conductors and insulators, in that most electrons are strongly bound, but a small number can be made to conduct. Certain semiconducting materials when excited by light will produce an electrical voltage, which can be used as a clean and renewable source of power.

A PV cell consists of adjoining layers of semiconducting material, usually of silicon doped with either phosphorus (n-type) or boron (p-type), although other materials can be used including gallium arsenide (see Boyle 1996 or Kazmerski 1997 for more details). The two different types, n-type and p-type, create a voltage between them. Incoming radiation in the form of sunlight can then provide an electron with the necessary energy to allow it to conduct. This leaves a positive charge or 'hole' in the semiconductor. If connected to an external circuit, the movement of electrons creates a current, which will flow under the influence of the voltage created.

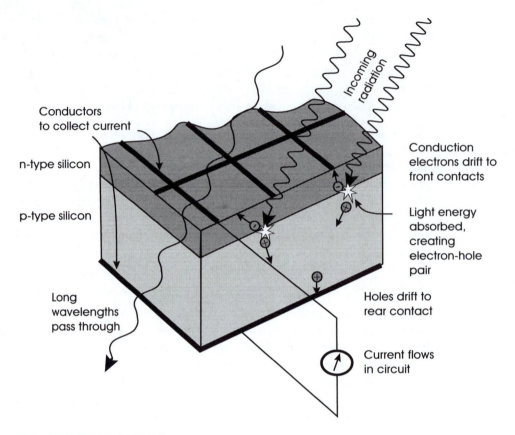

Conductors to collect current

n-type silicon

p-type silicon

Long wavelengths pass through

Incoming radiation

Conduction electrons drift to front contacts

Light energy absorbed, creating electron-hole pair

Holes drift to rear contact

Current flows in circuit

Figure 2.6 *Photovoltaic cell.*
Source: adapted from Boyle (1996) p. 98

Typically a single solar cell will be around 10 cm square and can produce a voltage of 0.5 V and a current of up to 2.5 A in full sunlight (insolation of about 1,000 W m^{-2}), giving peak power output of 1.25 W. The solar energy incident on this solar cell would be 10 W, of which 1.25 W is converted to electrical power – 12.5 per cent efficiency. The capacity of this cell would be termed 1.25 peak watts (W_p). A square metre of PV array would have 100 such cells giving capacity of 125 W_p. In actual use output would generally be lower due to dirt on the glass, ageing of components, higher temperatures, and the module not being perpendicular to the sun – in addition to which of course the sun does not always shine (see Box 2.3).

The efficiency of a PV module is limited by the quantum processes involved in the transfer of energy from the photon to an electron in the cell. One photon gives its energy to one electron as a single 'quantum' – there is no such thing as half a photon or half an electron. The energy of a photon in the incoming radiation depends on its frequency, and is given by $E = hf$ where h = Planck's constant = 6.6×10^{-34} J s (from Chapter 3, Equation (3.15)). In order to be absorbed in the

Box 2.3

Give me sunshine

Given full sunshine of 1,000 W m^{-2} and a 12 h day, the theoretical maximum total insolation per square metre would be:

$$I_{max} = 12\ h \times 1,000\ W$$

$$= 12\ kWh\ or\ 43.2\ MJ\ per\ day \quad (1\ kWh = 3.6\ MJ)$$

At 12.5 per cent efficiency this would produce output of 5.4 MJ or 1.5 kWh per day from a square metre of module, and if it could be produced all year that would make approximately 550 kWh.

A typical winter day in the UK has average insolation of around 1.5 MJ m^{-2} on a horizontal surface – less than a tenth of that in a typical June day of 18 MJ m^{-2}. Siting the PV module on a S-facing slope at a 45° angle increases the winter insolation figure to around 3 MJ m^{-2}. Over the year this equates to around 1,000 kWh m^{-2}. At 12.5 per cent efficiency, this would produce annual output of 125 kWh from a 1 m^2, 125 W$_p$ PV array. A hotter location such as California, Australia or southern Spain would have around 2,000 kWh m^{-2}, producing 250 kWh, while the highest insolation regions such as the Sahara reach around 2,500 kWh m^{-2}, which would produce 312.5 kWh from the same square metre of PV array. A 'tracking' device, moving such that it always faces the sun, would increase output further.

However in order to produce large amounts of power, large areas are always needed. For comparison, a 1 GW coal-fired power station would produce 8,760 GWh per year (if on all the time). The area of PV array needed to match this, in the Sahara, would be given by:

$$Area = \frac{8,760 \times 10^6}{312.5}$$

$$= 28 \times 10^6\ m^2 = 28\ km^2.$$

or a total area of 7 km by 4 km covered with PV modules, without allowing for gaps between modules, lower efficiency in real life or transmission losses – and an area 2.5 times greater if it were in the UK.

---

semiconducting silicon, the photon must have a minimum of 1.1 eV (where 1 eV is 1.6×10^{-19} J) of energy to make an electron conduct. A photon with less energy than this cannot be absorbed, while one with more energy than this only transfers 1.1 eV to the solar cell with the remainder being wasted as heat. A photon with energy of 1.1 eV would have a frequency of:

$$f = \frac{1.1 \times 1.6 \times 10^{-19}}{6.6 \times 10^{-34}}$$

$$= 2.7 \times 10^{14}\ Hz$$

Wavelength, $\lambda = c/f = 1.13 \times 10^{-6}$

This wavelength and frequency corresponds to infra-red light, quite close to the visible part of the spectrum. Thus any infra-red or other radiation of higher wavelength cannot be absorbed and is wasted, while visible light, UV and higher frequencies are absorbed but partially wasted as only 1.1 eV is converted to electrical power. For instance, a photon in the visible blue region with wavelength 4×10^{-7} m has energy of 3.1 eV, of which 2 eV will be wasted.

In theory, in normal sunlight maximum efficiency of 30 per cent could be achieved, due to these quantum energy effects. But these factors account for only part of the loss in efficiency – in addition a certain proportion of photons will be reflected at the glass, a few will pass right through, and then there are other losses in the resistances of the wires etc., resulting in actual efficiencies of 12–15 per cent.

PVs when in use have very little environmental impact aside from the visual and land-use issues, although in their production they do require considerable energy, and risk emissions of the various toxic elements such as arsenic used in manufacture. Currently PV power is used for a wide range of applications, including remote farmhouses, radio repeater stations, traffic monitoring, remote environmental monitoring equipment, health centres and vaccine refrigeration in developing countries, and satellites. In any situation where a grid connection is impractical or expensive and the power requirement is small, PVs (often used to charge batteries) may prove cheaper, more reliable and more convenient than the alternatives of batteries that need replacing or diesel generators. However for large-scale power production the costs are still high, although decreasing rapidly. One promising development is the use of PV modules as building components, in roofs or wall cladding. This reduces costs as it can be used in place of conventional materials, and also means no additional land is needed.

Energy storage

One disadvantage of many renewable energy forms is that they can be intermittent and unpredictable. Some, like solar or tidal, are predictable but intermittent (it always gets dark at night, the tide always follows a regular cycle), while others such as wind and wave are both, and some, like hydroelectric and geothermal, offer continuous and controllable power. Intermittency leads to problems in power supply as people understandably do not want the lights to go out if the wind is not blowing. A cheap form of energy storage would be an obvious solution to this problem, although the electricity grid system can cope with some of the variability.

The operation of the grid system has to cope with variability in demand from night to day and from summer to winter, while most power stations cannot be switched on and off rapidly. In particular nuclear stations must be run 24 hours a day whether the demand is there or not. This also leads to a need for storage, to make use of off-peak power. Peaks are met in a number of ways: by keeping power stations running well below capacity so they can rapidly be brought up to power; by peak power plants that are only turned on for a few hours each year and by pumped storage (pp. 62–3).

The grid can be used to incorporate unpredictable and intermittent supplies in the same way. For low penetration of renewables (up to 20 per cent or so of electricity), the variation is indistinguishable from the natural variation in demand and can be compensated by the same methods. If a lot of energy were generated from different renewables their variability would cancel out over the country to some extent: strong winds in Scotland may not coincide with those in Wales, while the tides in the Mersey are out of phase with those in the Severn. Hydro, biomass and geothermal could be operated as peaking plants when other sources were unavailable.

However for added flexibility, to reduce the need for excess capacity in the grid and for non-grid connected systems such as remote islands or farmhouses, a cheap, reliable means of storing electrical energy would be extremely useful. Unfortunately energy storage is generally technically difficult, inefficient and expensive. Possibilities for energy storage (as well as pumped storage) include:

- batteries;
- other chemical fuels such as hydrogen;
- heat stores;
- mechanically in flywheels; and
- as compressional energy in compressed gases.

While some of these are appropriate for certain situations – for instance flywheels can be used in buses as discussed on p. 76, and batteries are useful for provision of small amounts of power – none can store a large enough amount of energy at a low enough cost to be more widely applicable.

One promising technology to use electricity stored in chemical form is the fuel cell. Fuel cells are akin to batteries, except that they can be 'recharged' by supplying them with an external source of fuel. They work by oxidising a fuel, usually hydrogen, in the presence of a catalyst of platinum or nickel. The reaction produces water and energy in the form of an electric current, and so is virtually pollution free. Efficiencies are high as fuel cells do not involve the thermodynamic losses inherent in any heat engine, so 60 per cent efficiency is commonly achieved (Boyle 1996).

Fuel cells have been promoted as a key component of the 'hydrogen economy', where hydrogen would be produced from renewable energy sources then used in fuel cells to power vehicles, as an alternative to fossil fuels. They could also be used as a form of energy storage in grid systems, as they can be switched on and off virtually instantaneously.

An alternative catalyst of copper and cerium can be used to produce a fuel cell powered by methane (Seungdoo Park *et al.* 2000). In this case CO_2 is also produced and released, whereas with a nickel catalyst any carbon in the fuel would not react and clog up the catalyst. In addition to powering vehicles, this technology has the potential to act as a portable, clean and flexible combined heat and power plant. Homes or industries could install fuel cells to provide for both heating and electricity requirements powered by natural gas, with much greater efficiency than current gas-fired power stations.

Energy use in transport

Transport accounts for around 35 per cent of all energy used in industrialised countries, and this proportion is growing. Emissions including carbon dioxide (CO_2), carbon monoxide (CO), oxides of nitrogen (NO_x), volatile organic compounds (VOCs) and particulates are high, traffic produces pollutants where they are least wanted at street level, and emissions are more difficult to reduce or monitor than from large stationary sources such as power stations. While things are changing with the use of catalytic convertors and unleaded petrol, and manufacturers have made some improvements in efficiencies, transport still produces a huge amount of global warming CO_2, and relies on limited petroleum resources.

There are four main reasons cars need energy. By Newton's first law, no energy is required just to keep a mass moving in a straight line at a constant velocity. However energy is needed for:

- acceleration;
- going up hills;
- overcoming air resistance; and
- overcoming rolling resistance.

Acceleration produces kinetic energy and so is given by Equation (2.4), $KE = \frac{1}{2}mv^2$. As Newton's law specifies 'in a straight line', energy is needed to turn corners, to provide momentum in a different direction – this is why in a car you corner smoothly by reducing speed initially then accelerating into the curve. The energy used in acceleration ends up being dissipated by the brakes as heat when the vehicle slows down.

Going uphill requires potential energy, given by Equation (2.5), $PE = mgh$. While some of this energy may be regained freewheeling down the other side of the hill, most of it will be lost by braking.

Air resistance takes the form of a drag force (pp. 10–11), which has been shown experimentally to be approximately proportional to velocity. This relationship can be written as:

$$a = -kv \tag{2.13}$$

where k is the drag coefficient. The negative is because it is decelerating – air resistance slows down the car. So drag force is given by:

$$f_d = ma$$
$$= mkv \tag{2.14}$$

Energy used against this force is then given by Equation (2.2):

$$E_d = \text{force} \times \text{distance}$$
$$= mkvs \tag{2.15}$$

This shows that the energy use depends on velocity and distance.

This is an approximation that is satisfactory at normal speeds. In fact this drag force has two components – rolling resistance (mainly the friction of the tyres against the road) and air resistance. The rolling resistance is proportional to speed and the mass of the vehicle as shown, but the air resistance at higher speeds becomes proportional to the velocity squared (similar to the formula for raindrops given on p. 18), hence fuel consumption starts to rise rapidly above about 100 km h^{-1}. Air resistance is measured by the drag coefficient that depends on the shape of the vehicle, with sports cars having the best aerodynamics (minimising turbulence at high speed) and square boxes the worst – a Porsche 944 has around two-thirds the drag factor of a Lada Riva or a Mini.

Using the approximation given above, total energy required by the vehicle will be given by:

$$E_{tot} = \text{KE} + \text{PE} + E_d$$
$$= \tfrac{1}{2}mv^2 + mgh + mkvs \tag{2.16}$$

However the amount of energy in the form of fuel needed by the vehicle will be greater than this, because of the engine efficiency. Energy efficiency is typically only around $\eta = 18\%$ on average over a journey, i.e. only 18 per cent of the energy in the fuel is being converted to mechanical motion, with the rest being thermodynamic losses, dissipated as heat through the exhaust, cooling system and radiative losses. Cars are generally designed to have maximum engine efficiency at about 90 km h^{-1} (56 mph), due to convention and regulation, so petrol consumption will increase more rapidly past this speed, and at lower speeds more fuel is wasted in engine inefficiency. Engines are also very inefficient at low loads, using a considerable amount of fuel just to turn over in neutral. One innovative design for engine economy is a car whose engine switches off altogether whenever load is low – stopped at lights, coming down hills or cruising at moderate speed – and starts up again automatically when needed. While this has great potential for fuel saving, there are obvious technical and safety issues to do with starting again reliably and immediately.

For a long journey at constant speed without hills, the drag component will predominate and fuel use in litres per 100 km would be roughly proportional to speed. As motorway driving accounts for a high proportion of miles driven, improving aerodynamic design and reducing tyre resistance can cut fuel use considerably, as would a speed limit of 100 km h^{-1} or less.

On an urban cycle, with a lot of stopping and starting, the acceleration part will be more important, while in a hilly area at low speeds the work done against gravity would become important – in these cases the drag factor becomes less relevant, and small, light cars are better suited. The chief ways to reduce this consumption are reducing weight or improving engine efficiency.

As most energy from acceleration and hills ends up being dissipated by braking, one alternative is to somehow store this energy and use it to speed up again. A possible way of doing this is by use of a flywheel, which has been developed experimentally for urban buses. As buses are constantly stopping and starting and are heavy, this is an ideal application – up to 40 per cent of their energy is dissipated in braking. As the bus slows down, rather than conventional brakes it transfers its kinetic energy into speeding up a heavy flywheel to a very rapid rotational speed. The flywheel is then used to power the vehicle when it speeds up again, slowing down its rate of rotation. To be effective, the flywheel must not be too heavy, as this adds weight to the vehicle increasing its energy needs. To store enough energy therefore it must rotate very rapidly and have very low friction, which may require it to be encased in a vacuum as well as having extremely high quality, well-lubricated bearings. These systems can have a very significant impact on fuel economy, but also increase the cost of the vehicle significantly which is why they are not more widespread.

Energy efficiency of different transport modes

Although the above discussion centres on cars, the equations of course apply equally to trains, buses, bikes and even aircraft to some extent, although the relative importance of the four components – acceleration, work against gravity, air resistance and rolling resistance – may change. Table 2.4 compares energy use for each of these components for various transport modes.

Trains are fundamentally more energy efficient than cars in all four components. So a high speed train can travel at a much faster average speed than a car, and yet still be several times more energy efficient in terms of energy use per passenger mile.

Trains require some friction with the rail to allow them to brake, which results in energy use. An innovative rail system uses a magnetic monorail to support it on a very powerful magnetic field, matched by electromagnets in the train itself. This reduces rolling resistance virtually to zero, improving efficiency and reducing noise. As there are no wheels, the train now needs a different form of drive, which is

Table 2.4 *Energy use in various transport modes*

Energy used for	Train (inter-city)	Aeroplane	Ship
Acceleration	Moderate on long trips, as few corners and no need to start/stop	High on take-off due to high speed	Low
Gravity	Low as few hills	Very high	Zero
Air resistance	Moderate due to length	Moderate due to aerodynamic shape	Very low due to low speed
Rolling resistance	Moderate as on rails	Zero	Very low as on water
Comments		Most energy used in take-off	Low speed, limited by the speed of waves in its wash
Overall energy use	Low	Very high	Very low

provided by changing the magnetic field to move the train forward on magnetic forces.

Air travel is one of the most energy-intensive forms of transport available, because of the high speeds and heights at which they fly. Passenger aircraft fly at a high altitude (around 10,000 m), which not only avoids atmospheric disturbance from the weather and ground based obstacles, it also reduces air resistance because of the lower atmospheric pressure, and allows use of the jet-stream. Once at altitude, a level flightpath and low air resistance mean the aircraft can travel very fast without fuel use being prohibitively expensive. Unfortunately to get to this altitude requires doing a lot of work against gravity. Comparing the components in equation (2.16), for an aircraft to reach a cruising speed of 200 m s^{-1} (about 700 km h^{-1}) at 10,000 m would require 100,000 J kg^{-1} of PE and 20,000 J kg^{-1} of KE, expressed per kg of aircraft mass. This is some 300 times more per kg than for a car accelerating to 100 km h^{-1} at ground level, requiring only 385 J kg^{-1} KE and no PE. Because of this, weight is far more important in determining fuel use than in ground-based transport. Aircraft must be built from very lightweight aluminium alloys, and limit their payload or baggage allowances to keep weight down. The fuel needed for a long journey is itself very heavy – in fact the weight of fuel needed limits the maximum length of journeys. For instance a Boeing 747 flying from London to Hong Kong non-stop would leave with 200 t of fuel on board. The heavier the plane, the more fuel needed, and so the heavier the plane – a catch-22 situation.

By contrast, water transport combines minimal friction, low speed and travelling horizontally making it energy efficient and therefore cheap for moving bulky, heavy freight over long distances. Despite the growth of air freight, shipping is still the main bulk carrier for international trade. The main drawback is that ships are slow. The maximum speed of a boat is governed by the movement of waves in its wash, and hence is governed by its waterline length as the speed of a wave depends on its wavelength. The exceptions to this rule are craft that do break free of their bow and stern waves, such as speedboats, jet foils and hovercraft. By lifting up at one or both ends of the boat, they can avoid being held back by their bow-wave or sternwave and can exceed their waterline-length speed. This greatly increases their energy consumption however, being the maritime equivalent of an aeroplane 'breaking the sound barrier'.

Comparison of specific energy use

Comparing energy use in different transport forms is not always straightforward, as they differ in terms of speed and typical length of journey. Specific energy use can be expressed per tonne-km, or per passenger-km, representing the typical amount of energy used per km for each tonne of payload or for each passenger respectively. The latter measure is further complicated as it depends upon the number of passengers, often well below the maximum capacity – a full car is more efficient than an empty bus.

Table 2.5 gives figures for primary energy use for five freight modes (Royal Commission on Environmental Pollution 1995). These figures would be subject to considerable variation – in particular, fuel used in air transport depends strongly on the length of journey, making it far less efficient for short trips. It is clear that the potential for energy saving by switching freight from road to rail is large, while reducing air traffic would result in even larger savings. For those raw materials where it is possible such as oil and gas, pipelines provide the most energy efficient means.

Energy use is not the only, or arguably the most important, environmental impact of transport. Carbon dioxide (CO_2) emissions follow a similar pattern to energy use, but other air emissions including oxides of nitrogen (NO_x), volatile organic compounds (VOCs), carbon monoxide and sulphur depend upon the fuel used, combustion conditions and emission controls, so are not proportional to energy use. Other major concerns include safety, infrastructure and cost. If internal flights in the USA or Australia were replaced by car or train journeys, energy use would decrease, but there would be a need for more roads and railways to be built often in wild and remote areas, together with the ground-level development they would bring, which could have worse environmental impacts than aeroplanes flying overhead.

Table 2.5 *Energy use in freight transport*

Mode	Energy consumption kJ tonne-km
Road	2,890
Rail	677
Air	15,839
Water	423
Pipeline	168

Electric vehicles

Electric vehicles, or ZEVs (Zero-Emission Vehicles), have the potential to reduce the urban emissions from car traffic. Of course these vehicles do not really have zero emissions, unless they are powered by electricity from wind turbines or solar cells. But at least the emissions from power stations can be monitored and controlled, and unlike vehicle emissions, they are not produced at street level. ZEVs are available in California due to legislation to promote them from the regulatory body, requiring all motor manufacturers to provide them.

Apart from having no emissions in use, a ZEV is quiet and very efficient in energy use as electric motors can achieve efficiencies up to 90 per cent, compared with 18 per cent for an internal combustion engine. This also makes them very cheap to run. A major limitation in ZEV design is the amount of energy that can be stored in a battery. The weight of a conventional lead-acid battery is a major constraint on energy availability, limiting power and range, such that top speeds will be low and the vehicle will typically need recharging after 70 to 80 miles. Nickel metal hydride batteries increase the range and power as they have higher power density. To recharge takes several hours, typically done either at home overnight, or while at work or shopping, in a car park recharging station.

Petrol/electric hybrids are also being developed, using the petrol for long trips and electric in town with some of the advantages of both. Use of hydrogen fuel cells to power cars is another zero-emission option, as described on pp. 73–4.

Energy in the biosphere

Energy fixed by photosynthesis provides the basis of the food chains that support almost all life on Earth.

Photosynthesis

Photosynthesis is the process by which green plants, photosynthetic bacteria and algae convert the energy in sunlight together with carbon dioxide and water from the air into chemical energy in the form of carbohydrate in their leaves, which is then used to build the sugars, starches and proteins that make up plant matter.

Chlorophyll in the chloroplasts in leaves and algae absorbs energy from sunlight, part of which is used to split water molecules:

$$2H_2O + energy \rightarrow O_2 + 4H^+ + 4e^-$$

The oxygen produced escapes to the atmosphere, while the ionised hydrogen continues the process, carrying the energy. It passes along a chemical chain, until finally the H^+ effectively reacts with carbon dioxide to form carbohydrate, which uses the energy:

$$4H^+ + 2CO_2 \rightarrow 2[CH_2O]$$

The general form of carbohydrate, CH_2O, can then be built upon to form the many other compounds used in plants.

This reaction is reversed in respiration, the process by which all living things use energy – carbohydrates are broken down in the cell and combine with oxygen, releasing energy, water and carbon dioxide in the process. Combustion is also the reverse of photosynthesis – hydrocarbons in biomass and fossil fuels react with oxygen to release energy, water and carbon dioxide. The amount of carbon dioxide released in combustion per unit of energy will thus be equivalent to the amount initially taken up from the atmosphere during photosynthesis.

The rate at which photosynthesis occurs depends upon several physical factors including the irradiation level, temperature, water supply, supply of other nutrients and rate of diffusion of CO_2 into the chloroplasts. The photosynthetic rate can be modelled, as a function of all these variables. For instance, Figure 2.7 shows how the rate varies with available radiation at different values of CO_2 concentration and temperature. The term **photosynthetically active radiation** or PAR, is used to denote the amount of sunlight that can be used in radiation, as some is reflected, or of the wrong wavelengths – see Box 3.2. Under a specific set of conditions, one of the variables may be the limiting factor determining the rate, often the level of CO_2. For instance, if all the CO_2 that diffuses into the chloroplast is being used, no matter how brightly the sun shines the rate will not increase. Plants evolve to suit their environment, such that the rate is maximised under normal conditions. Features such as leaf shape and texture, orientation, thickness and size of the stomata (pores) regulate the water and CO_2 supply and help in temperature control. If the major nutrients (nitrogen, phosphorus and potassium) are in short supply they will limit the rate, as can minor nutrients and trace elements.

The effect of changes in the environment on growth rates can be difficult to predict, as the photosynthetic rate depends on so many factors. Plant growth can be stimulated by increasing the appropriate variable – by the application of fertilisers, watering, controlling external temperatures and so on. In some cases supplying one particular missing trace element in minute quantities can greatly increase growth, where this is the limiting factor. It has been suggested that fertilising the oceans with iron could increase primary productivity of phytoplankton, by removing an important limiting factor in their growth, thus fixing more CO_2 from the atmosphere and redressing the impact of these emissions on the climate.

Plant growth rates, climate and atmospheric carbon dioxide are closely linked, as plants are a major sink for CO_2 from the atmosphere. It has been found in trials that increased atmospheric CO_2 levels can lead to increased growth in many species, known as carbon fertilisation. This may mean that some regulation to CO_2 levels

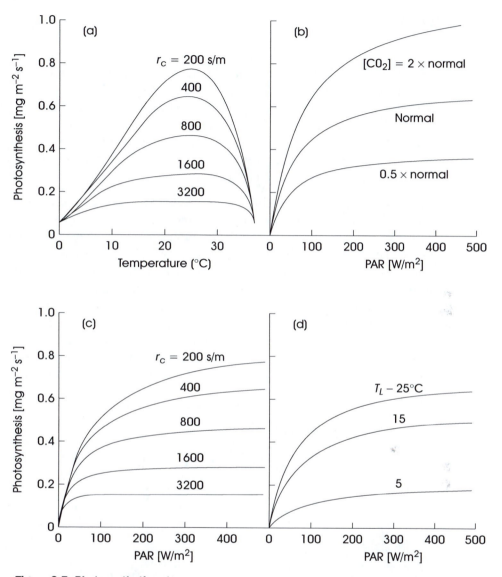

Figure 2.7 Photosynthetic rate.
Source: reproduced with permission from Campbell (1977) p. 118

will occur across the planet – as atmospheric concentrations increase, so will plant growth levels, increasing this carbon sink and reducing the rate at which CO_2 builds up in the atmosphere. However changes in climate will affect growth rates in other ways, for instance at increased temperatures respiration rates also increase, resulting in a lower net uptake of carbon dioxide. A changing climate will inevitably mean many natural species find they are no longer well adapted to their environment, and so growth may decline due to increased temperatures or reduced water availability.

Trophic levels

Within an ecosystem, energy and materials are passed from one species to another via a food chain. The food chain can be described in terms of **trophic levels**. The first trophic level is the **autotrophs**, or **primary producers**, which provide the basic input into the food chain by photosynthesis. The second trophic level eats the first, and the third level eats the second and so on. For instance grass photosynthesises (first trophic level) and is eaten by rabbits (second trophic level), which are then eaten by foxes (third trophic level) which then die and are fed on by decomposing bacteria (fourth trophic level).

Energy passes from each trophic level to the next with losses at each stage, as each organism uses energy to live, move around and so on. In terms of entropy, each successive level represents a lower entropy state, maintained by continual use of energy converting it to a high entropy form (low temperature waste heat). The amount of energy and biomass produced at the first trophic level is the **net primary production** of an ecosystem, which determines the amount at all higher levels. Only around 10 per cent of this available energy is passed to the next trophic level, the remainder being lost as waste heat. Thus the largest amount of energy production at any stage will be in the lowest trophic levels. There may however be more biomass at higher levels. For instance, at certain times of year zooplankton mass may exceed that of phytoplankton. This is because the

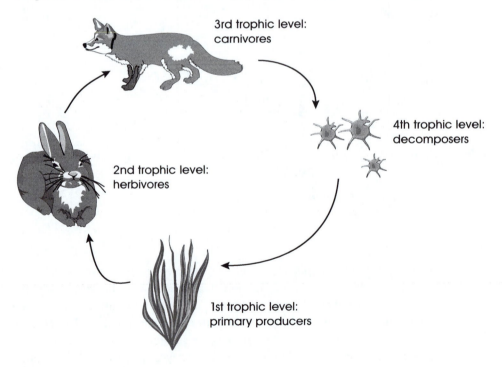

3rd trophic level: carnivores

4th trophic level: decomposers

2nd trophic level: herbivores

1st trophic level: primary producers

Figure 2.8 *Trophic levels.*

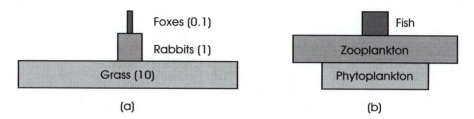

Figure 2.9 *Trophic energy pyramids: (a) annual energy flow at each trophic level (arbitrary units); (b) biomass present at each trophic level.*

phytoplankton is highly productive but the zooplankton is longer lived, and so the mass of phytoplankton at any one time may actually be less than the mass of zooplankton it is supporting.

A food web is a more realistic model, as many species live at more than one trophic level. For instance humans as omnivores eat green plants, putting us at the second trophic level, but also eat animals and fish from higher up the food chain. Likewise decomposing bacteria are not too fussy about whether they are acting upon a rabbit, a fox or a human.

These concepts are treated in more detail in Jones (1997).

Other biological energy sources

Photosynthesis is not the only source of energy for life. Some bacteria use other chemical sources of energy, for instance from hydrogen sulphide (H_2S). The chemosynthetic energy production is analogous to the photosynthetic process. In summary form, the process is:

$$CO_2 + H_2S + O_2 + H_2O \rightarrow H_2SO_4 + CH_2O$$

These chemosynthetic bacteria provide the energy for life purely from the energy in H_2S without the need for sunlight or green plants. Other forms exist that use other chemical substances as their energy source, including organic molecules such as hydrocarbons. In some cases these properties may be used in bioremediation of contaminated land. Specific micro-organisms can be introduced that consume a pollutant such as hydrocarbons in an oil spillage.

The sulphur oxidising bacteria mentioned above form the basis of complete ecosystems that are non-photosynthetic – the communities found around deep sea smokers or hydrothermal vents. These systems exist in the very deep oceans, where geothermal activity releases mineral rich water, containing H_2S, at high temperatures on the sea floor. The bacteria cluster around the vent, feeding on the H_2S, and provide a food source for organisms including worms, shellfish and other

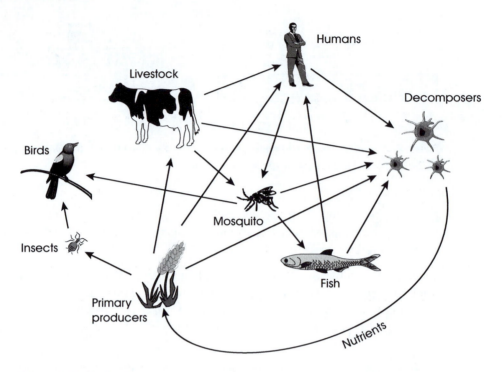

Figure 2.10 *Food web.*

larger creatures. The systems can contain hundreds and thousands of species not found elsewhere, all adapted to live in complete darkness under extreme pressure. The communities are self-sufficient, clustering around a vent for their food source, creating oases on an otherwise barren deep sea floor.

Another important class of chemosynthetic bacteria is the nitrogen fixers. The bacteria of the *Rhizobium* genus live in nodules on the roots of nitrogen fixing plants such as legumes. They live in a symbiotic relationship, in which the plant provides a home and energy supply for the bacteria, and they convert atmospheric nitrogen into a usable form for the plant to use in the formation of proteins needed for growth. Atmospheric nitrogen is in the form N_2, which is very stable and requires a considerable amount of energy to convert it into a usable form. Bacteria such as the photosynthetic cyanobacteria in the soil convert N_2 to a soluble form such as ammonia, which can then be converted by other bacteria into nitrites and nitrates that are usable by plants. In the *Rhizobium* bacteria, the energy source is provided by the plant in the form of sugars, which receives nitrates in return. This allows leguminous plants to grow on very poor, low nitrogen soils where little else could survive.

Biomass energy

Biomass energy in the form of fuelwood to supply human needs has been used since the first cavemen sat around their cooking fire, and is still the largest renewable energy form used in the world today. In some developing countries, fuelwood and other biomass energy such as straw, stalks and dung, used as the main fuels for cooking and heating, provide over 90 per cent of national primary energy use. Wood is also widely used in Europe and America, particularly in rural areas, while the production of energy from wastes and fuel crops is increasing.

While biomass fuels produce CO_2 in combustion, the amount produced is equivalent to that used in photosynthesis when they grow, and so they have the potential to form a sustainable, carbon-neutral energy resource. There are many other environmental issues to consider, including emissions such as particulates and NO_x, hazardous emissions including dioxins from incineration of certain wastes, and land use in energy crops. However it is now widely recognised that biomass could provide a valuable and cost-effective energy resource with much lower environmental impact than fossil fuel.

The energy content of biomass fuels is generally lower than their fossil fuel counterparts, as shown in Table 2.6. This is one factor that limits their value, together with their being more bulky, often heterogeneous in nature, and having a high water content. This makes them more expensive to transport and store, and less convenient to use, although new technologies can overcome these difficulties.

Biomass energy use is increasing in Europe in many new forms, using various organic wastes to produce electricity or heat, and collecting gas from landfill sites

Table 2.6 *Energy content of some biomass fuels*

Fuel	Typical energy content, MJ kg^{-1}
Wood (air dry)	15
Dung (dry)	15
Straw (depending on crop)	13–18
Coconut and groundnut shells, olive pits	20
Domestic refuse, unsorted	9
Paper and card	17
Industrial wastes (depending on industry)	7–20
Fossil fuels for comparison:	
Coal	29
Petroleum	42
Natural gas	55

Sources: Hall *et al.* (1982); Barnard and Kristofferson (1985); Boyle (1996); TERES (1994)

to use for energy production (TERES 1994 gives details). Energy can be produced from a wide range of industrial wastes from paper, textile, chemical or food industries to name but a few. For instance a famous tea bag manufacturer, when it converted its tea-bags from being square to being circular, calculated that the little bits of paper cut off to make the circular teabags could in theory be pelletized to provide enough energy to power the production line. Agricultural wastes such as straw and poultry litter also provide a huge energy resource which is now being used for electricity production in some cases. Liquid organic wastes, such as slurry and food processing wastes, can be fermented to produce biogas, which is mainly methane and similar to natural gas in its combustion.

Apart from using these waste materials, crops can be grown specifically for fuel. This can include coppiced trees for wood and various crops for liquid biofuels. Oilseeds such as rape, corn and sunflower can be converted into biodiesel to replace diesel fuel in vehicles. A variety of crops from sugar cane and corn to jerusalem artichokes can be fermented to produce ethanol, to replace petrol, used widely in the USA and Brazil. Ethanol production per hectare varies from as little as 250 l ha^{-1} produced from corn under less favourable conditions, up to 12,000 l ha^{-1} for sugar cane in tropical climates (Boyle 1996).

The overall efficiency of biomass in converting sunlight via photosynthesis into useful energy is very low. In Box 3.2, we will see that the efficiency of photosynthesis is limited to 5.5 per cent from physical considerations, that is a maximum of 5.5 per cent of the sunlight striking the leaf is converted into chemical energy in the plant. Under real conditions, it will be less than this. Much of the sunlight is lost because it does not fall during the growing season, or does not meet growing leaves but strikes soil, tree trunks, dead leaves etc. and is absorbed and re-radiated or reflected. Unless the plant is receiving optimal nutrition it may not be able to use all the available energy – in general the photosynthetic rate will be governed by a lack of nutrients or CO_2 and so operate at less than its maximum efficiency. This results in between 0.5 and 1 per cent of solar energy being converted into biomass energy in the plant, giving a yield per hectare of typically 5–10 t of dry biomass, depending on climate. More energy may be required to dry the fuel or to process it into a usable form. Efficiencies in use are also generally lower than for fossil fuels – varying from around 85 per cent for a state-of-the-art pelletised fuel boiler, down to just 5–10 per cent for a typical open fire.

Energy analysis shows that certain energy crops risk putting more energy into their production than can be produced as useful energy. Conventional agriculture in Europe and north America uses large inputs of energy, in the form of energy-rich nitrogenous fertiliser, use of tractors and other equipment, crop processing and transporting the final product often hundreds of miles to its final market. Most foodstuffs require more energy inputs than available as calories in the food. Typically, to produce 1 kCal of food energy uses 6 kCal in the form of these energy inputs, in addition to the solar input. While for food this is not a critical problem, (as one cannot eat coal), for biomass crops much less energy intensive

Box 2.4

How much energy in a day's human labour?

Let's say a manual labourer consumes a diet supplying 4,000 kCal of energy a day. The labourer spends 8 hours per day working, and uses 2/3 of her energy on work (the other third at rest or asleep), and has a metabolic efficiency of 40 per cent – fairly generous assumptions. 1 kCal = 4.2 kJ.

$$\text{Useful energy output} = \text{energy input} \times \text{efficiency}$$

$$= 4,000 \times 4.2 \times 2/3 \times 40\%$$

$$= 4,480 \text{ kJ}$$

$$= 4.48 \text{ MJ}$$

As 1 kWh = 3.6 MJ, this is equal to just about 1.2 kWh of energy – a few pence worth. No wonder machines have replaced manual labour in so many jobs! We have all become accustomed to using amounts of cheap, commercial energy that are vast compared to our own human energy.

farming must be practised for the process to be either economically or environmentally worthwhile.

Land required for biomass production is large but not totally prohibitive. Given that the total world fertile land area is 8.7 billion ha, with a production level of 5 t ha^{-1} of biomass with energy content 15 GJ t^{-1}, if this entire area were given over to energy production it could produce 6.5×10^{11} GJ of energy. This equates to about twice total world energy consumption. Thus all our energy needs could be provided from biomass grown on half the usable land area. This need not necessarily mean growing half as much food, as much of the bioenergy could be produced as by-products from agricultural production and forestry or from wastes after their primary use. Of course in Europe, with high population densities and high energy use, far more land would be needed than in Africa or South America. While providing 100 per cent of our energy from biomass is an implausible scenario, biomass could certainly play an important part in meeting our energy needs.

Summary

- The First Law of Thermodynamics describes the conservation of energy, which may be converted between different forms but never created or destroyed.

- The Second Law of Thermodynamics describes the tendency of entropy to always increase, linked to irreversibility, which together with the first law determines efficiencies of energy conversion processes.

- All energy can be defined either as kinetic energy (KE), due to movement, or potential energy (PE), due to an object's position.

- Renewable energy forms including water and wave power stem from solar power, and their potential may be assessed from an understanding of KE and PE.

- Transport uses energy for acceleration, work against gravity and work against friction including air resistance. Consideration of these factors demonstrates that air transport is inherently energy intensive, while rail and water are far more energy efficient.

- Photosynthesis in green plants forms the basis of the food chain, although some ecosystems survive on a non-photosynthetic basis.

- Biomass fuels can provide a sustainable substitute for fossil fuels, with no net carbon dioxide production.

Questions

1 A domestic gas fire has an efficiency of 80 per cent while an electric fire is 99 per cent efficient. Electricity from a coal-fired power station is generated at 35 per cent efficiency, with transmission losses of 1.5 per cent. Losses of gas in distribution amount to 1 per cent of supply. What are the overall efficiencies from gas to heat and from coal to heat?

2 What is the annual energy output from a 600 MW coal-fired power station, assuming it is available (switched on) 85 per cent of the time? Answer in J and kWh (or their multiples).

   Typical domestic electricity usage is 4 kWh per day. How many households can the station supply?

   The energy content of coal used in the station is 29 MJ kg^{-1}, with carbon content of 24.1 kg C GJ^{-1}. How much coal is burnt annually by it, and how much CO_2 is released (measured as carbon, assuming 35 per cent efficiency)?

3 The Sports Council estimate that a fit cyclist can use 600 Kcal h^{-1}, compared with 100 Kcal at rest.

   (a) What is the total power output of the cyclist?

   (b) If the efficiency of muscle at converting food into movement is 40 per cent and the cycle efficiency is 90 per cent, what is the useful motive cycle power?

   (c) How does this compare with the energy consumed by a moped getting 80 mpg? (energy content of petrol: 45MJ l^{-1}; 1 mile = 1.6 km; 1 gal = 4.5l)

4 In a small HEP station, water flows from a lake through a pipe to a turbine 80 m below the dam.

   (a) What is the potential energy of 1 m^3 of water at the top of the pipe?

   (b) If the overall station efficiency is 60 per cent and water flow rate is 1,000 m^3 hr^{-1}, what is the power output of the station?

5 A wind engineer is choosing between two sites. A windy hilltop is available with average wind speeds of 7 m s^{-1}, or a less exposed lowland site with average speeds of 5.5 m s^{-1}.

Due to planning restrictions, the developers have been told they must use a smaller turbine size on the hilltop than the lowland site. The larger size costs 60 per cent more for a turbine twice the size, but both have the same power coefficient and efficiency. Which site should the engineer recommend on cost grounds?

6 A car has mass 400 kg and drag coefficient 0.015.

(a) Calculate the energy needed to drive the car at 50 km h^{-1} constant speed over 100 m; and at 110 km h^{-1}.

(b) Calculate the KE gained by the car accelerating from 0–50 km h^{-1}, and from 0–110 km^{-1}.

(c) Calculate the PE of the car gained by climbing 100 m.

(d) How far would the car have to travel at constant 50 km h^{-1} to use the same amount of fuel as it used accelerating to that speed?

(e) Compare the answers to a, b and c. In a typical journey, which do you think uses most energy – drag, acceleration or hill climbing:

(i) in town?

(ii) on a motorway?

(iii) driving over a mountain pass with average gradient 1 in 5 (20 per cent)?

(f) Use the answer to (a) to find petrol consumption at constant 110 km h^{-1}, in miles per gallon, given energy content of petrol is 45 MJ l^{-1}, and assuming engine efficiency of 18 per cent. Why do cars not achieve this?

7 (a) A power station produces 600 MW which is transmitted at 132 kV. What is the current flowing?

(b) If it is transmitted through 100 km of transmission lines with a resistance of 3×10^{-3} Ω km^{-1}, what is the transmission loss in MW?

(c) What is this as a percentage of the electrical power produced?

Answers to numerical parts

1 electricity: 34.1%; gas: 79.2%

2 4.47 TWh or 1,240 TJ; 3,060,000 households; 122,000 t coal and 85,300 tC

3 (a) 0.7 kW   (b) 0.25 kW   (c) 28 kW

4 (a) 784 kJ   (b) 131 kW

6 (a) 8.3 kJ/18.3 kJ   (b) 37 kJ/187 kJ   (c) 392 kJ   (d) 445 m   (f) 124 mpg

7 (a) 4545 A   (b) 6.2 MW   (c) 1.03%

Further reading

Energy, Society and Environment. D. Elliot. Routledge, London, 1997. Covers technologies for energy production and use in society, with a sustainable development perspective.

The Future of Energy Use. R. Hill, P. O'Keefe and C. Snape. Earthscan, London, 1995. Discusses energy use mainly from a policy perspective but also with some technical details.

Renewable Energy: Power for a Sustainable Future. G. Boyle (ed.). Oxford University Press/Open University, Oxford, 1996. This OU text covers all the renewable sources in a very comprehensible and thorough volume.

Renewable Energy: Sources for Fuels and Electricity. T. B. Johanssen, H. Kelly, A. K. N. Reddy *et al.* (eds). Island Press, Washington DC, 1993. Comprehensive and slightly more technical than the OU book.

Environmental Biology. A. M. Jones. Routledge, London, 1997. Includes use of energy in the biosphere and trophic flows in Chapter 3.

An Introduction to Biophysics. G. S. Campbell. Springer Verlag, Heidelberg, 1977. Includes energy balances and energy use in the biosphere.

③ Heat and radiation

Heat is a form of energy that is present in all matter. It is important in human energy use, biologically and throughout the natural environment. The following key concepts are covered in this chapter:

- Heat is a form of energy, governed by the laws of thermodynamics
- Heat can be transmitted by convection, conduction and radiation, which describe important energy flows in both the built and natural environments
- Heat engines convert from heat to useful mechanical work, whose thermodynamic efficiency is constrained by their precise physical configuration
- The electromagnetic spectrum covers a wide range of radiation types, all with similar properties
- Many inaccessible environmental systems can be monitored by means of remote sensing, allowing sophisticated investigations given sufficient understanding of the properties of radiation

Heat forms a major part of energy consumption, with many opportunities for reducing energy use, and is also relevant biologically as all living things must balance their heat budgets to maintain temperature within a tolerable range. Radiant heat is a form of electromagnetic radiation, which covers the spectrum from X-rays to radio waves, all having certain properties in common. One application requiring an understanding of the properties of radiation is remote sensing, used in many environmental studies.

Heat and temperature

Heat is a measure of the thermal energy of vibrating molecules. The hotter a substance is, the more its molecules vibrate. As heat is a form of energy, it is measured in joules, and can be converted into other energy forms such as motion or electricity. It is governed by the First Law of Thermodynamics, which is the law of conservation of energy – that energy (as heat or other forms) can never be created or destroyed.

Temperature is a property of an object that determines the direction in which heat flows. According to the Second Law of Thermodynamics, when heat flows it will always go from a hotter object to a cooler one. Temperature and heat are linked, but not equivalent. Two objects at the same temperature may contain very different amounts of heat; and the addition of heat to an object does not always result in an increase in temperature.

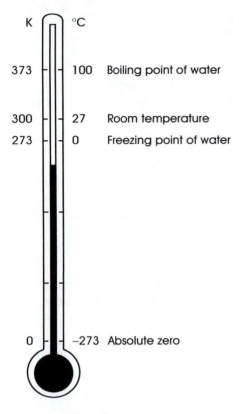

Figure 3.1 The Kelvin and Celsius scales of temperature.

In SI units, temperature is measured in Kelvin, written K (*not* '°K'), illustrated in Figure 3.1. The Kelvin scale starts at absolute zero – when molecules aren't vibrating, and it is impossible to get any colder. Absolute zero is 0 K and −273°C. There is no such thing as a minus temperature in Kelvin. As a degree Kelvin is equal to a degree Celsius, any temperature in Kelvin can be converted to Celsius by subtracting 273. Thus the freezing point of water is 273 K or 0° C; and room temperature is about 20° C or 293 K. In this book both are used, but Kelvin must be used for thermodynamics and the gas laws where absolute temperature is needed.

Heat capacity

The amount of heat energy needed to increase an object's temperature by a set amount depends upon its heat capacity. The heat capacity C of an object is the amount of heat needed per 1 K rise in temperature (unit: J K^{-1}), so the heat required, Q, for a rise of ΔT is:

$$Q = C \,\Delta T \qquad (3.1)$$

For a uniform substance, heat capacity is proportional to mass. The specific heat capacity, c, is the heat capacity per unit mass of a substance (unit: J kg^{-1} K^{-1}). For an object of mass m, heat capacity is $C = cm$ so Equation (3.1) now

$$Q = cm \,\Delta T \qquad (3.2)$$

Some specific heat capacities are shown in Table 3.1. High heat capacity means it takes a lot of energy to heat up the substance, and a long time to lose that energy in cooling down. Note that water is particularly high – in fact it is the highest of any common substance, which is why the oceans tend to keep temperatures stable compared to inland areas.

For instance, water has specific heat capacity of 4.2 kJ kg^{-1} K^{-1}. To raise the temperature of 1.5 kg of water by 80°C (when bringing the kettle to the boil for instance) would take $4.2 \times 1.5 \times 80 = 504$ kJ of heat energy. If the kettle has a rating of 2 kW, this is equivalent to 2 kJ sec^{-1} (as 1 W=1 J sec^{-1}), so the temperature rise in 1 second would be $2/(4.2 \times 1.5) = 0.32°$ C. The kettle will boil in $504/2 = 252$ s or 4 minutes 12 seconds – although in reality it would be longer

Table 3.1 *Specific heat capacities and latent heats*

Substance	Specific heat capacity C, KJ kg⁻¹ K⁻¹	Latent heat of fusion, KJ kg⁻¹	Latent heat of evaporation, KJ kg⁻¹
Water	4.2	334	2,260
Ice	2.1	334	–
Alcohol	2.5		867
Iron	0.47	205	
Aluminium	0.91	393	
Mercury	0.14	12.5	272
Carbon tetrachloride	0.84		193
Wax	2.9	176	
Wood (dry)	1.7	–	–
Stone	0.9	–	–

because of heat losses to the air. If the kettle were filled with pure alcohol with specific heat capacity of 2.5 kJ kg⁻¹ K⁻¹ it would need less heat and so temperature would increase more quickly.

Heat capacity is an important consideration for energy use in building design. A building with massive stone floors and walls will store a lot of heat, and so take a long time to cool down or heat up, keeping the dwelling at a constant temperature. Air holds relatively little heat, and is constantly moving in and out of any building for ventilation purposes. Typically all the air in a building changes at least once per hour – so most of the heat in a building is held in its fabric. High heat capacity reduces energy demand, as heat gained from the sun during the day keeps it warm at night. Conversely in hotter climes, the building will not overheat as rapidly during the day, reducing air-conditioning demand. Heat capacity can therefore be as important as insulation or draught-proofing in the energy efficiency of a building, although it has to be included when first built, unlike insulation that can be added later.

Latent heat

Matter exists principally in three states: solid, liquid and gas (pp. 136–7). **Latent heat** is the heat energy needed to bring about a change of state, from solid to liquid or from liquid to gas, without affecting temperature, as opposed to **sensible heat** discussed so far which results in a rise in temperature. Latent heat is important in many energy transfer processes, in particular in the atmosphere (see Chapter 5). Some substances will also go directly from solid to gas by sublimation, for instance carbon dioxide or 'dry ice'.

Table 3.1 gives values for latent heats of fusion (melting) and evaporation (boiling) for some substances – they vary widely, but are generally high compared with specific heat capacities. It takes a lot of energy to change state, as inter-molecular bonds must be broken, although this change is not accompanied by any change in temperature. The energy goes into changing the molecular bonds, so that for instance 1 kg of ice at 0°C takes 334 kJ to turn into 1 kg of water at 0°C. This energy will be released when the water freezes again.

Whenever water evaporates, it absorbs a lot of heat energy from its surroundings to do so – and when the vapour condenses to form droplets of water, the heat will be released again. Because the amount of latent heat is quite large, it represents a significant amount of energy 'stored' in water vapour. Power station cooling towers use this principle. In order to cool the hot water emerging from the turbines, after most of the energy has been used to make electricity, the water is sprayed into the cooling tower from near the top. Cool air is drawn in at the bottom, and as the water falls through the tower some of it evaporates, cooling it further. By the time it has fallen to the bottom the water is cool enough to be discharged into a river, while steam emerging from the top carries away waste heat to the atmosphere. The process is effective and economic, but still highly inefficient in energy or environmental terms. Far better to use that hot water to provide heating for nearby homes or commercial buildings in a district heating scheme, as is done commonly in some countries such as Denmark, and then cooling towers would be a thing of the past.

Refrigeration and heat pumps

Refrigerators and air conditioners use the exchange of latent heat in their circulating coolant fluid (or refrigerant) to operate. A compressor liquefies the refrigerant, then pumps it into the cooling circuit where it is allowed to expand and evaporate. As it does so, it takes in latent heat from its surroundings, cooling the body of the fridge. The gases are then pumped into the compressor to be condensed again, releasing heat that is lost via the heat exchangers at the back of the fridge, and the process repeats itself. The net effect is to extract heat from your food and dissipate it at the back of the fridge.

This may appear to break the Second Law of Thermodynamics – i.e. you have managed to move heat from a cold area to a warmer one. However this is not so, because you are inputting energy in the compression process, so it is not a closed system and entropy increases overall. Heat moves from cool food to colder refrigerant in the cooling element of the cycle; the refrigerant is then heated by compression and heat flows from warm refrigerant to the cooler air temperature in the heat exchangers. At no point has heat moved from cold to hot directly.

A refrigerant to be effective must have certain properties – it must have a large latent heat of evaporation and high specific heat capacity; have freezing and boiling points at appropriate temperatures; be a good conductor of heat; be chemically stable and be able to withstand repeated pumping, condensing and evaporation cycles for many years without degrading or reacting with the pipes and compressor. For these reasons chlorofluorocarbons (CFCs) have been found to be the best things to use – until banned for environmental reasons. Alternatives include hydrochlorofluorocarbons (HCFCs), butane or ammonia.

A novel design of heating system uses the same principle as a refrigerator in reverse, to make a very efficient method of heating – called a **heat pump**. If you imagine a refrigerator, with the door open pointing out of a window and the heat exchangers round the back being used to heat your home, you are just about there. In practice, the cooling part of the cycle uses some external environmental heat source at ambient temperatures, such as groundwater. The 'refrigerant' fluid cools the groundwater and warms itself up, is pumped into your home where it is compressed, releasing a lot of latent heat which is used in the heating system. The cold liquid is then pumped back outside to warm up again. The net effect is for all of the electrical energy used in the condenser to be converted to heat, plus the heat from the groundwater is 'pumped' into the heating system against the temperature gradient. Thus several times more heat energy can be produced than the electricity put in, making the process very efficient indeed compared with conventional central heating systems.

Thermal expansion

When any substance is heated it will expand, due to increased vibration of molecules. This effect gives rise to the need for expansion joints in railway lines and bridges, to allow for temperature changes. Over large temperature changes expansion can give rise to stresses within materials, so for instance in large power stations the boilers must be heated up or cooled down slowly enough to allow for gentle expansion to prevent cracking. This is particularly important in nuclear reactors, where stress fractures from expansion could result in leakage or failure of the pressurised cooling systems.

At normal temperatures, as a reasonable approximation, this expansion can be assumed to be proportional to temperature change. Expansion is thus given by:

$$e = \lambda l (T_2 - T_1) \tag{3.3}$$

where e is expansion, l is original length, T_1 and T_2 are the temperatures before and after; and λ is the coefficient of linear expansion of the substance.

The coefficient λ will vary with temperature, increasing for most substances. At room temperature, λ for iron is $17 \times 10^{-6} \, \text{K}^{-1}$. In other words, a bar of iron 1 m long will expand by $17 \times 10^{-6} \, \text{m}$, or 17 μm, for each degree rise in temperature.

For liquids, the equivalent measure is cubic expansivity γ, as we are interested in the increase in volume of the liquid.

$$V_2 - V_1 = \gamma V_1 (T_2 - T_1) \tag{3.4}$$

where V_1 and V_2 are volumes before and after the temperature change – so the change in volume is given by the cubic expansivity multiplied by change in temperature.

For mercury, thermal expansivity is 1.82×10^{-4} at 0–100°C and is fairly constant over normal temperatures, allowing its use in thermometers. The thermometer can be calibrated with a linear scale – all the ticks are the same distance apart – as expansion is linear, a set amount for every degree.

Expansion is not always linear – for many substances, expansivity varies with temperature. Water is unusual in that it expands when it freezes. In fact water is most dense at 4°C, so as temperature rises from zero to 4°C expansivity is negative. It then rises from 0.5 at 5–10°C; it is 1.5×10^{-4} at 10–20°C, to about 6×10^{-4} at 80°C. A thermometer filled with water would have ticks four times as far apart at 80°C as at 10°C, and would be useless below about 10°C, as water at 5°C or 3°C would have the same volume. Plus of course below freezing the glass would break – overall you can see why mercury is used.

Expansion of water can be illustrated by the graphs in Figure 3.2, showing density and volume expansivity of water over a range of temperatures. As density is inversely proportional to volume, expansion reduces the density correspondingly

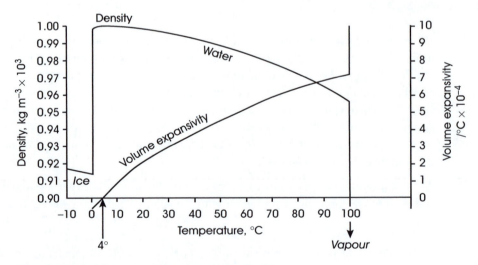

Figure 3.2 *Density and volume expansivity of water.*

(mass remains the same). Expansivity is the rate of change of volume with respect to temperature – in other words, the gradient of a graph of volume against temperature, and so is also related to the slope of the graph of density against temperature, as illustrated.

Water's variable expansion gives rise to complex effects in the ocean. The calculation of the expansion of sea water due to global warming is far from straightforward, as water temperature varies over the Earth and with the depth underwater. Also, water at less than 4°C will float, as its density is lower than that of water at 4°C. The bottom of deep oceans therefore will always be at 4°C, but the surface will be warmer in most places but colder or frozen near the poles. Predicting changes in ocean temperatures involves modelling of the transport of heat by convection to deeper levels, which depends on density, due to variations in temperature and also variations in salt concentration (connected with ice formation and melting).

The exceptional heat characteristics of water – high heat capacity and latent heat, variable volume expansion and expansion when frozen – together with its universal solvent properties make it an ideal basis for life. Underwater, the environment has very stable temperatures and all the available chemical building blocks in solution ready for reaction into life-forming proteins. Without water, any possible life-form would be inconceivably different from our own. The search for extra-terrestrial life has therefore concentrated on looking for planets that may contain water in a liquid form.

Transmission of heat

Heat can be transmitted by three distinct processes: convection, conduction or radiation.

Convection is the movement of liquids and gases brought about by changes in temperature affecting density and creating currents.

Conduction is the movement of heat through a substance, by exchange of thermal energy between neighbouring atoms, which can occur in solids, liquids and gases.

Radiation is the transfer of heat energy via electromagnetic (infra-red) waves, which are emitted from every object, as long as it is at a temperature above absolute zero.

A central heating radiator demonstrates all three heat transmission methods (Figure 3.3). Heat is first conducted from the water inside through the metal. The hot air surrounding the radiator will rise by convection, drawing in cold air at floor level from the rest of the room. The hot metal also heats the room and its occupants by radiating directly on them.

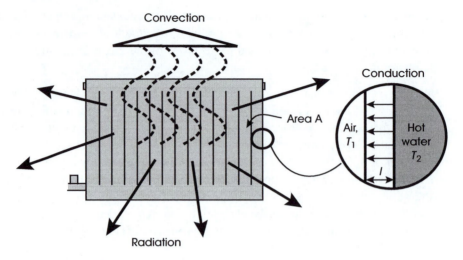

Figure 3.3 *Conduction, convection and radiation of heat from a radiator.*

Convection

Convection is important in climatology, as convection currents are the root cause of ocean currents and winds, which at a global scale serve to redistribute heat from the tropics to the poles. It is distinct from other heat transfer methods, as it relies on physical movement of warm material, and so can only take place in fluids (i.e. liquids or gases, that can flow). As warm air (or water) expands, its density decreases, and so it becomes lighter than colder air surrounding it, and tends to rise. Cold air is then brought in to take its place, which can lead to cyclic motion.

Convection currents also exist within the Earth itself (see also Box 4.1), with flows of molten rock moving around the fluid parts of the Earth's core, implicated in the production of the Earth's magnetic field. Nearer the surface, convection in the upper mantle is important in tectonic crustal movements and in the formation of volcanoes. The heat transfer in these flows can be massive, and they are significant in determining how fast the Earth cools.

Forced convection is the term for cooling or heating something by directing a current at it, for instance the wind-chill effect of wind cooling your body. This increases the rate of heat exchange over normal convection, but involves an external energy source.

Conduction

Conduction is most important in solids, as it is the only means of heat transfer in solids, while for gases or liquids convection will usually predominate. Conduction of heat is analogous to a stream flowing downhill. The water flow rate will depend upon how steep the stream is, along with its size and how rough the stream bed. In the case of heat, the amount conducted depends on the temperature gradient, the size of the conductor, and the conductivity of the material.

Temperature gradient is the difference in temperature divided by the distance it is being conducted. Heat flow is proportional to this temperature difference, and to area as a conductor with a large area will conduct more. The relationship is given by:

$$Q = \frac{kA(T_2 - T_1)}{l} \tag{3.5}$$

where Q is heat conducted per second (in Watts); k is the conductivity of the material, measured in W m^{-1} K^{-1}, A and l are area and length; T_1 and T_2 are the temperatures on each side (Figure 3.3).

When heat is conducted, energy is being passed from one molecule to its neighbours, primarily via the circulating electrons. When an atom is heated it vibrates more strongly, passing energy to the electron cloud which then passes it on to neighbouring atoms, making them vibrate too. Thus the materials with highest conductivities are those where molecules are close together, bound by interactions between their electrons (i.e. solids rather than gases) and those where the electrons are mobile between different atoms, such as metals. Metals are good conductors, while materials such as plastics or wood are poor conductors. Gases and liquids have low conductivities, but convey heat by convection as well.

Table 3.2 *Thermal conductivities of some materials*

Substance	Conductivity, W m^{-1} K^{-1}
Iron	76
Aluminium	210
Asbestos	0.13
Plate glass	1.1
Brick	0.13
Slate	2.0
Cotton wool	0.025
Ice	2.1
Water (10°C)	0.62
Paraffin Oil (0°C)	0.13
Air (0°C)	0.024

Insulating materials like fibre-glass or woolly jumpers rely on incorporating air into their structures, that has a very low conductivity. If this air is held in a closed-cell foam, as in expanded polystyrene, this will minimise heat movements from convection, so these are the best insulators.

Conduction is a relatively slow process, compared with convection or radiation. This leads to time lags in warming, as the heat at the surface is gradually conducted into the object. This is particularly relevant in climate, as the Earth and the oceans warm up very gradually with any increase in surface air temperature, as the heat gradually conducts downwards. It is this time lag that explains why afternoons are warmer than mornings, and why August is hotter than April (in the northern hemisphere) despite the sun being equally intense in both months.

A further parameter governing heat flow is the **thermal diffusivity**. If a material is subject to heat input that varies over time, thermal diffusivity describes how the heat 'spreads out' in between the changes in heat input. For instance, the sun shines on the ocean during the day but not at night. During the day, the surface layers heat up and heat starts to move down through the water by conduction. At night, the surface cools down, and the heat gained during the day continues to diffuse both down and up into the colder surface (neglecting convection for the moment). The rate of these heat flows, and hence how deep the changes in temperature at the surface penetrate, depends on the thermal diffusivity. The same applies to heating of soils by the sun, or to cooling of buildings at night. Thermal diffusivity is equal to $k/\rho c$, where k is thermal conductivity, ρ is density and c is specific heat capacity. In soils, it depends on both soil type and water content, as the chief components of soil – water, air, organic matter, quartz and clay minerals – have very different densities, heat capacities and thermal conductivities. Hence the soil type will determine temperature and the magnitude of variations in temperature at depth, between day and night (diurnal variation) or summer and winter. For a typical moist clay the thermal diffusivity will be around 5×10^{-7} m^2 s^{-1}, but this can vary from 1×10^{-7} m^2 s^{-1} for a dry peat soil to 8.5×10^7 m^2 s^{-1} for moist sand (Monteith and Unsworth 1990).

It is possible to define a **damping depth**, the depth to which surface temperature variations penetrate. This depends on the time period over which temperatures fluctuate. Damping depth can be expressed as

$$D = \sqrt{(\kappa\tau/\pi)} \tag{3.6}$$

where κ is thermal diffusivity and τ is the time period of the temperature variation.

For a moist soil with $\kappa = 5 \times 10^{-7}$ m^2 s^{-1} and for the diurnal variation in temperature $\tau = 24$ hours $= 86,400$ seconds, this gives a damping depth of 0.12 m. For annual variations, the time period is 1 year or 3.15×10^7 seconds, giving a damping depth of 2.24 m (Campbell 1977). Below 0.12 m, the soil will be expected to retain the average daily temperature, while below 2.24 m the temperature will remain at the average annual temperature. In arctic regions, this annual damping depth corresponds to the permafrost level. Wherever annual average temperature is below freezing, the soil below about 2 m underground will be permanently frozen.

Heat in buildings

Space heating in buildings accounts for about 30 per cent of UK energy use, so reducing these demands could have a big impact on CO_2 and other emissions. It is estimated that measures that are currently cost-effective, such as improved insulation and draught-proofing, could reduce energy use by 15 per cent, while far higher savings could be made by technically feasible but currently uneconomic measures. In hotter climates where air-conditioning is common, more energy can be used in cooling than heating, particularly in offices. Cooling energy demand can also be reduced by similar measures, insulating to keep heat out rather than in.

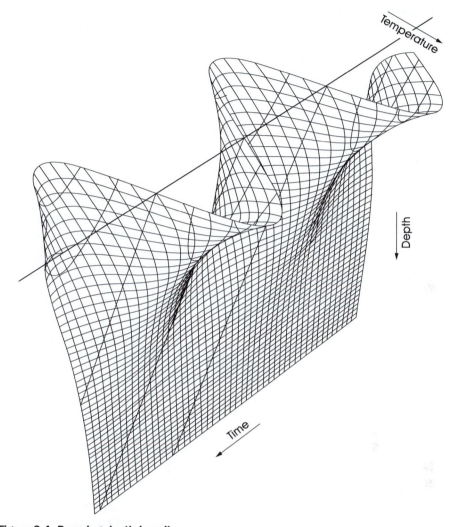

Figure 3.4 *Damping depth in soil.*
Source: reproduced by permission from Campbell (1977) p. 17

The comfort level experienced by the inhabitants of a building depends on several factors other than temperature. Draughts act to transfer heat away from the body, resulting in feeling cooler at the same temperature. Convective or fan heating, requiring air movements, will be less effective for this reason. Radiative heat warms the inhabitants directly, so one may feel warm in front of a fire despite low air temperatures. Thus warm (radiating) walls and a lack of draughts result in warm inhabitants without excessive air temperatures and so is more efficient. The presence of damp in buildings increases heat demand and heat losses, while high humidity makes hot temperatures less bearable.

In a building, a corrugated iron roof will conduct away far more heat than one of slate; while heat loss through windows will be more than that through walls. These

Table 3.3 *Common U-values for building materials*

Component		U-value, W K^{-1} m^{-2}
Solid brickwork	105 mm	3.3
	220 mm	2.3
	335 mm	1.7
Cavity brick wall	260 mm	1.19
	insulated	0.27
Corrugated sheeting	5 mm	5.3
Concrete	150 mm	3.4
Glazing	6 mm single	5.6
	2 mm airspace double	2.9
Roofs	uninsulated flat	3.0
	uninsulated 35° pitch	1.5
	with 100 mm glass fibre	0.35
Solid floor		0.3

Source: values from O'Callaghan (1992)

differences are due to the thickness of the material and the heat transfer properties of the materials involved. Insulation in buildings is often measured by a U-value, which is the amount of heat lost per unit area of wall, roof, window etc., allowing for its thickness and for multiple layers. The overall amount of heat loss depends on the conductivities of all the layers, and also on the rates of heat transfer between them, and heat transfer by convection (e.g. within an uninsulated cavity wall) and radiation away from its surface. The U-value combines these factors to give the amount of heat conducted through the wall in W K^{-1} m^{-2}, giving heat loss in watts per degree Kelvin difference in temperature between inside and outside, per square metre of wall area.

The amount of heat lost, which we will call Q_U, can then be estimated by summing the heat losses through all components – walls, doors, windows, roof and floors:

$$Q_U = \sum_i U_i A_i \Delta T \tag{3.7}$$

where U_i and A_i are the U-value and area respectively of the ith component, and ΔT is the temperature difference between inside and out.

Insulation only tells part of the story however, as much of the energy loss from a building is through ventilation. A ventilation rate of at least one air change per hour is needed for fresh air for the occupants, but in most buildings it is considerably higher, up to ten changes per hour. This comes naturally from air infiltrating through small cracks around windows and doors, whenever doors open, and through pores in bricks or other building materials. The heat lost in the ventilation air, Q_V, will be given by:

$$Q_V = \Delta T \, V R \rho_{air} C_{air} \tag{3.8}$$

where Q_V is heat loss per hour; ΔT is temperature difference, V is building volume, R is ventilation rate in air changes per hour, ρ_{air} is the density of air and C_{air} is the heat capacity of air.

Also important in the energy needed to heat a building are the sundry, or incidental, heat gains. These are the energy gains from heat lost by appliances, lights, people and any other energy using equipment in the building. They can be large, especially in buildings such as theatres with a high occupancy, or office blocks with a lot of

Table 3.4 *Standardised degree-days for the UK*

Month	Heating			Cooling		
	18.5°C	15.5°C	10°C	15.5°C	5°C	−20°C
January	488	395	226	0	17	705
February	426	342	189	0	23	652
March	390	297	134	0	64	803
April	319	233	96	5	114	837
May	235	151	39	14	192	963
June	148	77	9	26	265	1,015
July	88	42	4	96	380	1,155
August	100	45	5	57	338	1,113
September	162	83	10	14	245	995
October	268	177	48	1	158	925
November	359	275	124	0	53	719
December	439	346	176	0	29	755

Source: Vilnis Vesma Degree Days Direct, www.vesma.com 8.3.00

electrical equipment such as computers, copiers and so on. In a well insulated modern office, these sundry gains can be sufficient to provide all the heat needed for the building. Dependent on building design, heat from the Sun incident on windows can significantly reduce heating demand, but may also increase cooling demand.

Heating engineers simplify the calculation of heating demand by the use of degree-days. As external temperatures vary continuously, it is not practical to use the difference in temperature in estimating heat demand. Degree-days use average daily temperatures over years to represent monthly heating demand, under the assumption that the 'heating season' occurs when external temperatures are less than a baseline figure, usually 15.5°C. Degree-days are calculated by taking the amount temperature is less than 15.5°C, summed over all days in a month that temperatures are less than 15.5°C, in other words the heating degrees multiplied by the days.

An equivalent measure, cooling degree-days, is used to estimate air-conditioning and refrigeration demands. Growing degree-days can also be defined, used to estimate plant growth rates as a function of temperatures. Degree-day figures are calculated by the relevant national or state authorities and published as averages over several years. Table 3.4 shows standardised figures for the UK, for both heating and cooling, for various base temperatures.

Using degree-days, and entering the numerical values for density and heat capacity of air, Equations (3.7) and (3.8) simplify to:

$$Q_U = D \sum_i U_i A_i \times 24/1{,}000 \qquad (3.9)$$

$$Q_V = 1/3 \, DVR \qquad (3.10)$$

where Q is the heat demand in kWh and D is the degree-days figure.

From the law of conservation of energy, a building in equilibrium must have a heat balance, such that energy input, from the heating system, sundry gains and solar gains, is equal to output, via the fabric of the building and the ventilation rate:

$$Q_{in} + Q_{sundry} + Q_{solar} = Q_U + Q_V \qquad (3.11)$$

These separate elements can be estimated in an energy audit process, to find the areas with greatest potential for efficiency measures.

In the future, climate change will lead to changes in degree-day figures, which can be predicted although with inevitable margin of uncertainty. These can be used to study the impact of higher temperatures on heating and cooling demand.

Heat balance in animals and plants

Heat budgets in animals and plants are a key factor in their evolution. Warm-blooded animals have to keep their bodies at around 33–38°C, while cold-blooded ones don't mind cooling down but cannot freeze solid, and can only function actively within a similar temperature range. Fish and birds also need to regulate heat balances. Fur and feathers provide insulation, while sweating, panting and water/mud baths can be used for cooling. The colour of an animal's coat or skin determines heat gains from solar radiation, and also radiative heat losses as these depend on emissivity and absorptivity (see pp. 122–5). Plants too need to stay within a viable temperature range, both for survival and to maximise photo-synthesis. Their foliage size and shape may reflect this need in different climates.

A heat budget can be calculated for an animal, or for a plant, similar to that for a building. The heat balance equation averaged per unit surface area, has the form:

$$R + M = C + \lambda E + G \qquad (3.12)$$

where: R is net radiative heat gain, M is metabolic heat gain, C is convective heat loss, λE is latent heat loss and G is loss of heat by conduction to the environment (which is generally negligible, with the exception of underwater organisms). Each of these heat fluxes is expressed in W m^{-2}, heat per unit surface area. For any organism, each of these terms can be quantified, to model the fluxes of heat at the surface as functions of intrinsic and environmental variables. The equality must hold, due to the conservation of energy, hence the temperature will adjust to keep the balance – at higher temperatures, heat losses from radiation, convection and conduction all increase. To avoid temperature changes outside their survival range, the organism must control the factors affecting heat fluxes. Modelling heat balance allows the study of the conditions needed for the organism to survive and the adaptations it may make to external heat stress.

For a leaf, metabolic heat loss is minimal, so temperature depends chiefly upon incident radiation, evaporative and convective heat losses, which will vary with wind speed and humidity. The temperature of the leaf may be above or below air temperature, dependent upon the gains from radiation and the losses from latent heat, which closely links temperature to water use. A plant may adapt its leaf size and orientation to control radiative gains, and the size and number of stomata (leaf pores) can be varied to optimise evaporative losses. Stomata may close under conditions of drought and at night to reduce evaporative losses, affecting temperature, while waxy leaf coatings reduce evaporative losses further. Wilting is one strategy to prevent overheating under conditions of insufficient water supply – your pot of basil may perk up again very rapidly when it is watered. Plants adapted to hot, dry conditions cannot readily control temperature by evaporative losses. For instance Australian gum trees have many features to keep temperatures down without using water, such as waxy cuticles, pendulous leaves to avoid direct radiation in the heat of the day, and whitish coloured leaves to reflect radiation – some even shed leaves during the hot summer. However a plant has relatively few measures at its disposal to actively respond to heat stress, and temperature related adaptations must be balanced against the need to maintain water supply, light input and gaseous exchange systems needed for photosynthesis. Plants rely upon growing in a climate suited to their species, in terms of water supply, temperature, wind and solar input. Being immobile there are major risks from climate change, as entire forests and ecosystems could suffer from heat stress if temperatures rose past some critical level.

In an animal, heat losses are via conduction through the tissues and coat followed by radiation and convection from its surface; latent heat from panting and sweating, and heat in exhaled air. Heat gains are from metabolic activity, and from direct and diffuse solar radiation. The heat balance depends crucially on temperatures at the coat surface and the skin surface, while core body temperature must be kept to within a survivable range dependent on species. Some factors that determine heat loss, such as metabolic rate, body temperature and heat transfer characteristics are intrinsic to the animal, while others such as external temperature, solar irradiation and wind are environmental variables. Any animal has evolved to thrive under certain environmental conditions, and will survive over a range of conditions close to its optimum. This range is known as the organism's **climate space** or environmental niche. This space represents the range of temperatures, radiation and other external factors that the animal can survive over the long term, subject to their ability to vary intrinsic heat loss factors to cope with short-term variations. The survival of any one species in a given region will depend upon it being within the animal's climate space. To thrive, a much narrower range of conditions will be needed, towards the centre of the climate space, in which the animal expends the

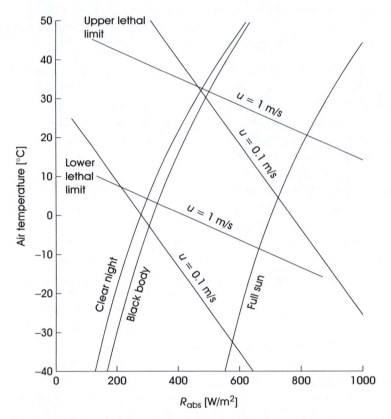

Figure 3.5 Climate diagram for a masked shrew.
Source: reproduced with permission from Campbell (1977) p. 90 (see text for explanation)

minimum of energy on temperature control – for humans this is termed the comfort zone.

In Figure 3.5, R_{abs} is radiation absorbed (from sun, surroundings etc.), which can vary between the lines for full sun and clear night, with the 'black body' line approximating a burrow. Windspeed u takes two values to give the upper and lower lethal limits of temperature for each speed, according to radiation absorbed. The enclosed area is the climate space – outside these limits the shrew will overheat or become hypothermic and would not survive.

To balance the heat budget, warm-blooded animals chiefly vary their metabolic rate, while cold-blooded animals vary body temperature. Vasodilation and vasoconstriction in surface blood supply are used to control heat flow to the skin surface, as reducing surface temperature reduces heat loss. Body temperature is reduced in the extremities in many animals including humans, to reduce heat loss while retaining core temperature. Some birds such as penguins have 'heat exchanger' mechanisms in the circulatory systems of their legs, such that hot arterial blood entering the legs is cooled by cold blood in the veins returning to the body – the same is true to a lesser extent in humans. Flamingos living in lakes fed from hot geothermal springs in Africa, such as Lake Nakuru, may actually use this mechanism in reverse to prevent the very hot water they wade in overheating their bodies. In fish, which are cold-blooded, body temperature is highly dependent on water temperature, resulting in some arctic species needing biochemical 'anti-freeze' to prevent their blood freezing.

There are many other adaptations and strategies to control temperatures across the animal kingdom. Fluffing up one's fur or feathers (the origin of 'goose bumps') and warmer winter coats can vary coat insulation. Hibernation involves a reduction in body temperature to reduce heat loss, together with a slowing of the metabolism to reduce the need for food. Burrowing animals can avoid extremes of temperature at the surface – as seen on p. 100 on the damping depth, the diurnal temperature variation only penetrates a few centimetres into the soil, while to avoid winter cold a deeper burrow would be required. The burrow also serves as shelter from wind, rain or sun; reduces radiative losses; can be insulated with straw and fluff, and it gives protection from predators, providing a safe and thermally stable home for young. The female polar bear rears her young cubs in a snow hole through the Arctic winter – kept at close to freezing point inside from the bears' body heat, this provides a far warmer environment than $-20°C$ plus wind-chill outside.

Evaporative heat losses may be the dominant form of temperature control in a hot environment, via sweating and evaporation through the lungs. A human can produce up to 1 kg per hour of sweat, resulting in heat loss per unit surface area of 375 W m^{-2} under conditions where all this evaporates. This is around seven times the basal metabolic heat production. If water is not available for these evaporative losses, in a dry environment such as the desert, dehydration and heat stress will result. An animal's resistance to heat stress will depend in part upon the reserves of water it is able to carry.

As heat loss by convection and radiation depends on the surface area of the animal, it could be expected that the metabolic rate would depend on surface area, which is experimentally found to be the case. Basal metabolic rates of around 50 W m^{-2} have been measured for mammals of all sizes from mice to elephants. Water reserves by contrast will depend upon body volume, so resistance to heat stress is characterised by the surface area to volume ratio. Large bodies and spheres/cubes have low surface area to volume ratios, while small volumed objects and long spindly legs have large ratios. So small creatures may have more difficulty keeping cool in hot, dry climates. Insects cannot use evaporative cooling, as they are too small to carry the necessary water reserves. However this simple relationship is far more complex in reality as other factors come into play, for instance small animals are less exposed to wind, reducing heat loss.

Humans go one step further in temperature control by using clothes and buildings, greatly increasing our ability to adapt to different climates, which was probably a major factor in the spread of the human race across the world. The human climate space now includes anything found on the planet, as we can adapt the environment to suit ourselves rather than the other way around.

Engines and thermal power production

The conversion of heat from fossil fuels into useful work requires engines, whether in cars or as turbines in electricity production. The physical features of these engines, including their size, temperature of combustion and efficiency, determine emissions of air pollutants such as SO_x, NO_x and CO_2, which contribute to major environmental problems. Following the discussion of engines and conventional power plant, some thermal renewable sources are described, namely geothermal, solar water heaters and passive solar.

Engines

The conversion from heat into useful mechanical energy is performed by a heat engine. In the simplest steam engine, the burning fuel boils water to expand when it forms steam, which then drives a piston. This basic concept lies behind all modern engines and thermal power generation.

In an internal combustion engine, fuel (petroleum) mixed with air is introduced into the base of a cylinder containing the piston. Ignition by a spark causes expansion of the air and exhaust gases, driving out the piston. As the volume of gas in the cylinder expands, it does some work, so energy is extracted from the system. The crankshaft acts to return the piston, expelling exhaust gases, ready to repeat the cycle. In a diesel engine, the principle is the same as in a petrol engine, but there is no spark: diesel fuel and air are compressed together, heating them to flashpoint so that combustion is spontaneous in the cylinder. The higher temperature and other details of the cycle make this thermodynamically a more efficient process.

For an internal combustion engine to run smoothly, the fuel must burn evenly and rapidly in the cylinder, requiring air and fuel to be well mixed and combustion to be complete, so that all products of combustion can be expelled as exhaust gases. This requires a high-quality, homogeneous fuel in liquid or gaseous form. The engine may be run on alternatives to fossil fuels, such as ethanol or biodiesel. Natural gas, propane or hydrogen may also be used in diesel engines with some modifications.

An alternative to the internal combustion engine is the Stirling engine. This uses two interlinked pistons within a single cylinder with one end being cooled and the other heated by an external heat source. The result is an engine that can be highly efficient, with its continuous power input and low heat losses, reaching close to its theoretical Carnot efficiency (pp. 110–11). In addition, as combustion is external to the cylinder, the fuel could be almost anything – wood, sawdust or briquettes of compressed domestic waste for instance. Unfortunately the engine design requires very high engineering tolerances to reduce gas leakage in order to work efficiently, making it expensive. However there is a lot of interest in producing a commercially viable Stirling engine, to produce power efficiently from these biomass sources, whether for low emission vehicles in Europe, small-scale combined heat and power production, or for electricity production in remote areas of the developing world.

Jet engines work upon a different principle. Rather than using the expansion of gas to drive a piston, that then drives a crankshaft to turn the wheels, the gases are expelled through a turbine, turning a propeller directly. In a plane this drives the craft forward and draws air in to keep the fuel burning continuously. Jet engines are inefficient and noisy, but also very powerful and compact.

Thermal power stations

The majority of thermal power stations – whether fuelled by coal, gas, oil, nuclear fuel or chicken litter – make use of steam turbines to convert from heat into mechanical energy. The turbine rotates a magnet in a magnetic field to produce electricity in the generating set (pp. 58–60).

Heat from combustion of the fuel, or from the nuclear reactor, is used to heat water to steam, which is then heated to a temperature as high as feasible given the materials involved, between 300 and 600°C. Steam at high temperature and pressure is fed to the turbine where it turns a propeller-bladed shaft. As the energy is extracted from the steam it cools and pressure reduces, continuing through a series of propeller blades of increasing size to extract the maximum possible work from the steam. It finally leaves the turbine to be cooled via cooling towers and expelled to the environment.

An alternative type of plant is the gas turbine, which is similar to a jet engine in its operation. Using oil or gas as fuel, the turbine blades are turned directly by the hot exhaust gases from combustion. Gas turbines are less efficient than steam turbines, but have the advantage that they can be fired up very rapidly and so are used for peaking plant.

A relatively recent development that combines the advantages of both technologies is the combined cycle gas turbine or CCGT. Fuelled by natural gas, these plants use both a gas turbine and a steam turbine. The gas turbine can start up rapidly, while feeding waste heat back into the steam turbine, greatly improving efficiency to around 50 per cent overall, with a consequent reduction in emissions per kWh produced. Their main limitation is that they require gas or oil as fuel.

The Carnot cycle

Why are engines and thermal power stations so inefficient? The efficiency of a process is limited by entropy, and a theoretical maximum thermodynamic efficiency can be calculated. A French engineer, Sadi Carnot, devised a formula to do just this in 1824. He devised a theoretical thermodynamic concept known as the **Carnot cycle**, in which a gas is heated, expands to do some useful work, cools and contracts again, as in a modern power station or engine. Carnot studied the changes in heat, energy and entropy, devising the maximum useful work that could be output without breaking the first or second laws of thermodynamics.

The physical changes occurring in an engine can be examined by use of two diagrams – the pressure-volume (pV) diagram and the temperature-entropy diagram (TS) diagram (Figure 3.6). The idealised case is the Carnot cycle, which produces the maximum useful work and thus the highest efficiency. From point 1→2 the gas is compressed, increasing pressure and reducing volume while entropy is constant. With heat input of Q_h the gas expands at constant temperature from 2→3 increasing entropy, followed by further expansion at constant entropy from 3→4 when temperature drops. It is then compressed at constant temperature, as heat Q_c is extracted producing net useful work of $Q_h - Q_c$, as the cycle returns from 4→1 to the initial conditions. Efficiency is given by the ratio of useful work to heat input, calculated from:

$$\eta = \frac{Q_h - Q_c}{Q_h} = \frac{T_h - T_c}{T_h} = 1 - \frac{T_c}{T_h} \tag{3.13}$$

where Q_h and Q_c are the heat input and heat output, while T_h and T_c are the temperatures at the hot and cold end of the cycle. In any real engine, efficiency will be lower as these changes will not be as distinct – temperature and entropy will not

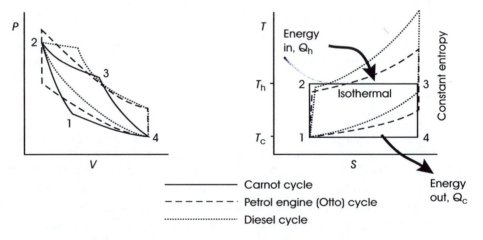

Figure 3.6 *TS and pV diagrams for a Carnot cycle and other engines.*
Source: adapted from Boeker and van Grondelle (1999) pp. 88–93

Box 3.1

Efficiency of a power station

For a typical coal-fired power station, the steam reaches about 500°C = 773 K: this is T_h. The cold temperature, T_c, is not less than ambient temperature: it can be taken as 300 K (27°C). Carnot efficiency is given by Equation (3.13):

$$\eta = 1 - 300/773$$

$$= 0.61$$

This is the theoretical limit, and once other losses are accounted for, most power stations end up with overall efficiency around 35 per cent. The remaining energy is lost as low grade heat and is generally dissipated by cooling towers, or into a lake, river or the sea.

For a nuclear station, temperatures are cooler, to restrict corrosion for safety reasons. At T_h = 300°C = 573 K, the Carnot efficiency will be:

$$\eta = 1 - 300/573$$

$$= 0.52$$

Again there are heat losses elsewhere, giving overall efficiency of 25–30 per cent. Thus the waste heat of 65–75 per cent of heat input contains between two and three times as much energy as the electricity produced.

---

be exactly constant, and the details determine the efficiency of the engine. The amount of useful work extracted in a cycle is low compared to the heat energy in the gas, so losses from friction will be proportionately greater. Actual efficiency is usually 50–60 per cent of the Carnot limit.

For a power station, where the cold temperature is limited by ambient temperatures, this means the higher the temperature of the combustion, the higher the efficiency, as shown in Box 3.1. However very high combustion temperatures are restricted by engineering and economic realities, as materials corrode more rapidly and distort at very high temperatures. High temperatures also increase emissions of NO_x, produced from nitrogen in the air. In addition to the thermodynamic losses there will be losses elsewhere, in the form of heat loss from the boilers, heat exchangers, efficiency of the generators and in transmission of electric power.

Geothermal power

Geothermal power is the use of heat from the subterranean regions of the Earth, either to generate electricity or to use directly as heat, often by industry or district heating schemes. In Iceland, geothermal energy is used to heat greenhouses to grow food in the cold climate, while in New Zealand it is used for both electricity and domestic heating in many areas. Historically the Maori people in New Zealand have used geothermal 'hotpots' for cooking as well as bathing in hot springs and heating their homes geothermally for hundreds of years.

While the central parts of the Earth have very high temperatures, they are totally inaccessible in practical terms. On average, the continental crust is 30 km thick, which is far deeper than drilling technology could achieve due to the phenomenal pressures and forces at these depths. The Earth is gradually cooling, at a rate of 0.06 W m^{-2}, with a temperature gradient of around 30°C km^{-1} near the surface. While these figures suggest the geothermal resource to be virtually useless, they represent averages, and there are many sites with much higher heat flows and temperature gradients.

There are two classes of geothermal region: geologically active and inactive regions. Active regions include areas of plate tectonic and volcanic activity near the plate margins, which may include geysers or outflows of magma. Temperature gradients are generally over 80°C km^{-1}. The inactive regions are those usually close to plate boundaries, with lower thermal gradients than active regions, in the range 40°C km^{-1} to 80°C km^{-1}. They include quite localised hotspots that may be due to a thin area of crust, the presence of heat-producing radioactive material (uranium or thorium), an area where the mantle temperature is particularly high or the heat conductivity of crustal rock is low.

There are two alternative geothermal technologies – use of aquifers, or use of 'Hot Dry Rock' (HDR). Almost all existing geothermal uses heat from aquifers: deep, high temperature water sources, usually consisting of brine and hence highly corrosive and noxious.

Hot Dry Rock (HDR) technology does not rely on the existence of an aquifer. An HDR plant first drills down to 5–7 km depth, then fractures the rock by a series of controlled explosions. Cold water is then pumped down to permeate the hot rocks, and hot water extracted for power production. This presents more technical difficulties than using an aquifer, and is more expensive, but allows a wider range of sites to be used.

Once the hot water or steam is brought to the surface, in either case it can be converted to electricity through a modified steam turbine and generating set, as for fossil fuel.

Geothermal energy has several environmental impacts, some of which can be reduced. It is not renewable, as the temperature of the aquifer or rock used will gradually decline as heat is extracted. The brine in aquifers can contain many contaminants as it comes from so deep down, including arsenic, mercury and other heavy metals. This problem is minimised as long as cooled brine is reinjected into the aquifer, which also extends the heat reservoir life. Steam contains gases, some of which are removed by passing through a condenser, but non-condensible gases can present a problem, especially hydrogen sulphide which gives geothermal plants their characteristic smell of bad eggs. Geothermal plants have also been associated with subsidence, and create noise, steam and visual impact from pipelines.

Solar water heaters

Solar energy is limited in its usefulness for space heating, as by definition most solar energy is available when least heat is needed – during the summer. For instance in the UK, solar irradiation in December is around 1.7 MJ m^{-2}, compared with 18 MJ m^{-2} in June. However for water heating, this is less of a problem as households use hot water all year round. The flat plate solar water heater provides an emission-free and often cost-effective source of energy for domestic hot water or larger scale uses such as heating swimming pools.

A solar water heater consists essentially of a series of pipes, coloured black, encased in an insulated box with a glass top (see Figure 3.7). The module is positioned facing the sun at an angle equal to the latitude to maximise solar irradiance. The pipes are incorporated into a metal sheet to increase their absorption, similar to a domestic radiator. The glass allows light to enter but helps prevent loss of infra-red radiation, while selective coatings on the glass improve efficiency further. In the simplest systems, a hot water tank is built into the top of the heater. These make use of the thermosyphon principle – hot water rises by convection to the tank, drawing cooler water through the heating element. They are common in many hot climates in countries such as Greece.

In cooler climates, this simple design cannot be used because of the risk of frost damage in winter. A more complex design uses anti-freeze in a closed cycle

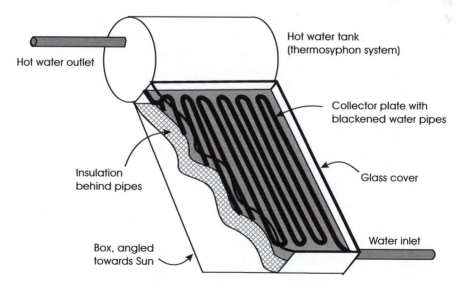

Figure 3.7 *The flat plate solar water heater.*

circulation system, connected via a heat exchanger to an insulated hot water tank inside the home. This increases costs and reduces efficiency.

Efficiencies are around 40 per cent, implying an annual output of around 400 kWh m^{-2} in the UK, where annual solar irradiation is 1,000 kWh m^{-2} approximately. A typical household would use 3–5 m^2 to supply hot water. Temperatures achieved are suitable for domestic hot water in the summer months, while in spring and autumn the system can be used to pre-heat water, combined with a fossil-fuelled system.

Passive solar design

Passive solar architecture consists of reducing heating demand by designing buildings such that their solar gain is maximised, particularly during the winter months, and their heat capacity is high enough to allow this heat to be stored during the night when outside temperatures fall.

A solar dwelling uses large areas of glass, facing south (or north in the southern hemisphere) to let in solar radiation, which warms the interior. Heat is then retained by the building's high heat capacity, low ventilation rate and good insulation. Insulation includes double or triple glazing and above normal levels of cavity wall and roof insulation, to prevent heat loss. Heat capacity is increased either by massive stone or concrete walls and floors, or by purpose built heat stores, which may include water tanks. Vertical windows, rather than horizontal rooflights, maximise solar gain in the winter months when the Sun is low in the sky and its input is most needed, while shades or blinds can prevent overheating in the summer. Glass has the useful property that it allows visible light to enter but it is less transparent to infra-red radiation, so heat cannot escape at the same rate. Externally, the building is oriented to face the sun and not be overshadowed by trees or neighbouring buildings, and may also have the main living rooms on the sunny side with bathrooms, hallways, stairs etc. on the north. Figure 3.8 shows an illustrative passive solar building.

By these means heat demand in buildings can be reduced significantly – indeed in many passive solar buildings there is no demand for additional heat input, with sufficient warmth being generated from the solar features, occupants and incidental gains from other electrical equipment such as lights and appliances. Unlike solar water heating moreover, passive solar is just as useful in cold climates as hot, because of the greater demand for space heating during spring and autumn when there is still a lot of useful solar radiation.

Passive solar is particularly appropriate at altitude, where night-time air temperatures plummet because the thinner atmosphere does not reduce radiative heat loss in the way it does at sea level. For instance in Ladakh in northern India, a specialist centre for solar energy has successfully encouraged passive solar for the

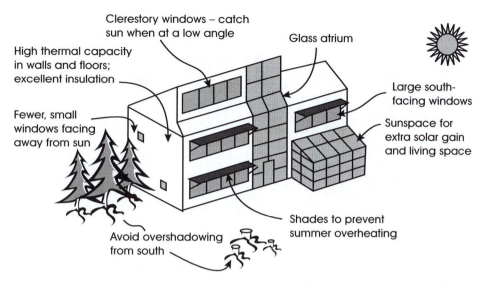

Figure 3.8 *Passive solar design.*

region. Ladakh is 3–4,000 m above sea level in the Himalayas, cut off by snow for nine months of the year and with night-time temperatures falling to −20°C, yet it is on the same latitude as the Middle East and hence receives a similar amount of solar radiation. Use of passive solar technology has greatly improved comfort standards in homes where it is used, while reducing the environmental problems caused by use of fuelwood.

It is remarkable that typically around a third of western energy consumption – and hence global warming CO_2 and other emissions – goes into heating buildings, when in theory all building heat demand could be supplied either by passive solar, by increasing insulation, or by district heating schemes utilising waste heat from power stations, with no additional emissions at all.

Radiation: the electromagnetic spectrum

Radiant heat is a form of electromagnetic radiation, like light, X-rays, radio waves and others. These are all electromagnetic (EM) waves, and have many properties in common.

An electromagnetic wave is a self-propagating system consisting of an electric field that induces a magnetic field at right angles to itself, which induces an electric field, and so on. Because the fields do not need any medium in which to exist, light and all other electromagnetic waves can travel across space. However in some ways they also represent particles – indivisible 'packets' of information which are called photons. To understand the details would require a knowledge of quantum

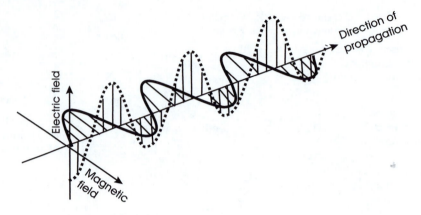

Figure 3.9 *An electromagnetic wave.*

mechanics and of relativity, but almost everything important about EM waves can be well understood without this.

In common with other waves, they have a wavelength, λ, measured in metres, and a frequency, f, measured in Hz where 1 Hz is one cycle per second (Figure 3.9). The different colours are light with different wavelengths – white light is a mixture of light of all wavelengths.

Although divided into the different bands, the spectrum is continuous – radiation of any wavelength is possible, and the values given are indicative only. Heat radiation is in the infra-red (IR) range.

All these waves travel at the same speed: the speed of light, which is constant in a vacuum. The speed of light is $c = 3 \times 10^8$ m s^{-1}. Frequency, wavelength and the speed of light are connected by a simple relation:

$$f\lambda = c \tag{3.14}$$

As c is constant, the shorter the wavelength, the higher the frequency and vice-versa. Knowing either frequency or wavelength, the other can be calculated – either one characterises the wave. For instance, a mobile phone uses the microwave frequency 1,800 MHz. The wavelength is given by:

$$f\lambda = 3 \times 10^8$$
$$\lambda = 3 \times 10^8 / 1,800 \times 10^6 = 0.167 \text{ m}$$

Table 3.5 *The electromagnetic spectrum*

	γ-rays, X-rays	Ultra-violet (UV)	Visible light: Blue ... Red		Infra-red (IR)	Microwave and radar	Radio waves SW ... FM ... MW ... LW			
λ, m	10^{-12}	10^{-8}	4×10^{-7}	7×10^{-7}	10^{-6}–10^{-4}	$10^{-3} - 0.1$	0.5	1	10	1,000
f, Hz	10^{20}	10^{16}	10^{14}		10^{13}	1–100 GHz	500 Mhz to 500 KHz			

Box 3.2

Wavelengths used by photosynthesis

Photosynthesis in green plants and other organisms was discussed on pp. 79–81. While the essential reaction of photosynthesis is the same in all of these, the biological details of the process differ in how light is captured, resulting in differences in the wavelengths of light used.

In green plants, photosynthesis makes use of light across the visible part of the spectrum to power the chemical reactions in plants. Light is captured by the pigments, chlorophyll-a and carotenoids such as β-carotene in the chloroplasts. These capture red and blue light, but not green, which is why leaves are green – it is reflected. The exact mix and nature of the pigments determines the leaf colour and the amount of light usable by the plant. The amount of light of the usable wavelengths incident on the plant is known as the photosynthetically active radiation or PAR. As the atmosphere is most transparent to visible light, this is by far the dominant type of radiation energy available, explaining why plants evolved to use this part of the spectrum. However PAR only accounts for about 50 per cent of total radiation.

Certain photosynthetic bacteria contain other pigments in addition to, or in place of, chlorophyll, allowing them to make use of other wavelengths. Cyanobacteria contain a mix of pigments called phycobilisome, containing phycoerythrin, phycocyanin and allophycocyanin, which each absorbs light of different shades including green, producing different characteristic colours. Other bacteria absorb in the near infra-red part of the spectrum, with pigments bacteriochlorophyll-a and bacteriochlorophyll-b, allowing them to thrive in environments where there is little or no visible light.

The wavelengths at which light is absorbed determine the efficiency of photosynthesis, in a similar way to photovoltaics (pp. 69–72). In both cases the quantum nature of light determines the amount of energy available. A photon of a certain wavelength contains an amount of energy given by Equation (3.15), $E = hf$. For a photon at the red end of the visible spectrum, with wavelength $\lambda = 700$ nm, this would mean it has energy of $E = hc/\lambda = 0.284 \times 10^{-18}$ J.

This photon excites a chlorophyll molecule, which can then use the energy to produce a free electron, used in splitting water and in producing carbohydrate. It is found that ten electrons are needed in the reaction per molecule of water or carbon dioxide, and produce chemical energy of 0.78×10^{-18} J. The efficiency is given by:

$$\eta = \frac{E_{out}}{E_{in}}$$

$$= \frac{0.78 \times 10^{-18}}{10 \times 0.284 \times 10^{-18}}$$

$$= 0.27$$

$$= 27\%$$

Actual efficiency is less than this, as a photon from the blue end of the spectrum, of wavelength 400 nm, would have almost twice as much energy input for the same output, halving efficiency. Only certain wavelengths can be absorbed so some photons are unusable, and some are reflected at the leaf's surface or absorbed by other tissues. Also around 40 per cent of photosynthetic production is lost through respiration, the plant's energy use just to stay alive. These considerations reduce the efficiency to around 5.5 per cent of normal sunlight on the leaf being convertible to chemical energy. In farming systems, the observed ratio of annual output to annual solar input is actually around 0.5–1 per cent.

Each photon carries a discrete 'packet' of energy. This is due to the basic principle of quantum theory, that small things come in discrete lumps rather than continuous amounts. The amount of energy carried by one photon is given by Planck's law:

$$E = hf \tag{3.15}$$

where E is energy, f is frequency and h is Planck's constant, h $= 6.626 \times 10^{-34}$ J s.

Although the amount of energy carried by one photon is small, in daylight there are zillions and quadrillions of photons, so it can add up to a lot. But in any interaction between radiation and matter, the amount carried by each photon does matter. One photon interacts with one atom, and needs a certain amount of energy for the interaction to take place.

From Planck's law, the amount of energy carried is proportional to frequency of radiation. So the most dangerous radioactive γ-rays and X-rays carry most energy, while radio waves carry least, and visible light is somewhere in the middle.

Transmission, absorption and reflection of radiation

For light, a material may be transparent, translucent or opaque depending on how much light it lets through. The same is true for other types of radiation. However a substance need not behave the same way for all types. For instance, walls are opaque to visible light, but transparent to radio waves. When a substance is opaque, the radiation is absorbed, and heats up the substance. Light is absorbed by the Earth, which heats it up.

For most substances, some frequencies go through and some are absorbed, with the amount absorbed shown by a material's **absorption spectrum** over a frequency range. Figure 3.10 shows the absorption spectrum for water. On the peaks, radiation of these frequencies is absorbed by water; where there is a gap or a 'window', radiation passes through.

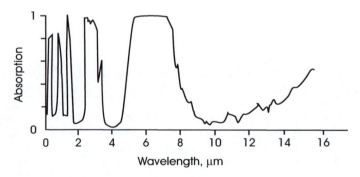

Figure 3.10 *Absorption spectrum for water in the infra-red region.*

Box 3.3

Animal and bird vision

Colour vision depends on cells in the eye that are sensitive to different wavelengths of light. In the human eye, these cells are known as cone cells, due to their shape, sensitive to red, blue and green light. Depending on the actual wavelength of light seen, the different colours of cones will be stimulated to different levels, and the combination gives the brain a signal for a broad range of colours. In addition the eye contains rod cells, which are more sensitive, but do not differentiate between the colours. At night, the world goes monochrome, when there is no longer enough light to stimulate your cones.

The concept of primary colours is due to these three colours of cone cell. Physically, the spectrum is continuous and there is no such thing as a primary colour. Colours between the primaries are seen as a mix of primary colours. For instance, the blue cones are most sensitive to light with a wavelength of 4×10^{-7} m, while the green cones are sensitive to 5.5×10^{-7} m. Light with a wavelength of 5×10^{-7} m stimulates both and so is seen as turquoise. However if the eye is shown a mixture of light at 4 and 5.5×10^{-7} m, it will respond identically with a turquoise signal. This is the principal of a TV screen – the eye can be tricked by displaying combinations of red, green and blue light dots into thinking it is seeing the full spectrum.

The vision of different animals varies considerably. Some such as moles or deep sea fish have lost their vision entirely, as they live in the dark and have no need of it. Many animals do not have colour vision, but only see in monochrome because they do not have cone cells. Others who are more dependent on their vision have a wider range of colours visible than mankind. For instance many birds can see into the UV part of the spectrum, invisible to us. This must make their view of the world very different. The colours of birds' plumage are designed to look best including UV reflections, with many birds having distinctive UV markings, only visible to us in computer-enhanced photographic images. Many birds have iridescent feathers, such as starlings and magpies. These do not have a pigment, but due to their surface they reflect light such that it interferes at certain wavelengths, bringing out many colours like in bubbles or an oil film on water. They may look fairly drab to us, unless caught in the right light. But again in UV light, the iridescence is greatly increased and the plumage looks much brighter and more colourful.

It has also been shown that UV is used in hunting by some hawks preying on grassland mammals such as shrews. The shrews can hide themselves in the grass and are fairly well camouflaged which would make them difficult to spot. However they run along set trails, which they scent mark with their urine. Urine contains chemicals that reflect UV – so the hawk can actually see their trails, and follow them to catch their prey.

A microwave oven works because the microwaves interact with water in food at a frequency that makes water molecules vibrate (or strictly speaking, spin) and heat up. The microwaves are of a frequency dominant in water's absorption spectrum. Fortunately most food contains enough water for this to work, with other components conducting heat away from pockets of water, while your plate will stay cool. Ice has a different absorption spectrum to water. When you use a microwave to defrost food, it is much less effective as less is absorbed – until the ice crystals start to melt and tiny pockets of water form, which then heat up and melt the surrounding ice.

Every atom or molecule has its own unique absorption spectrum, or series of frequencies that are absorbed. This together with the emission spectrum, the range of frequencies emitted, provides a 'signature' for any substance, useful for environmental analysis (see pp. 126–7).

From many surfaces, radiation is also reflected, with the proportion reflected determined by its **reflectivity**. Light is reflected in a mirror, and similarly heat or UV would also be reflected. In an ordinary mirror, maybe 99 per cent of light is reflected and 1 per cent absorbed. A white painted wall might reflect 75 per cent of light striking it, while a fresh snowfield might reflect 90 per cent. A wall painted black absorbs most of the light, with only perhaps 10 per cent reflected. A surface need not have the same reflectivity for different wavelengths. Blue paint reflects blue light, but absorbs other colours' wavelengths and so looks blue.

Thus radiation may be transmitted, absorbed and reflected from any object. By the law of conservation of energy, total radiation energy incident is equal to that absorbed, plus that transmitted, plus that reflected. The amount of each depends on the characteristics of the object (i.e. its absorptivity and reflectivity, which determine colour) and the wavelength of radiation.

Other wave properties in EM radiation

EM radiation shares certain properties with other waves, including refraction, interference, diffraction, and scattering, which were explained in general terms on pp. 31–3.

When light enters water (or any other transparent medium) at an angle it is refracted – instead of continuing in a straight line, it is deflected at the water–air boundary. The angle through which the light is refracted varies with its wavelength. Because of this difference with wavelength, the refracted rays will be split out into

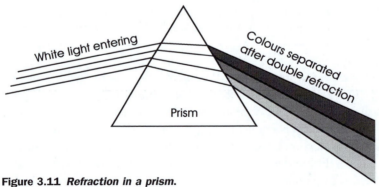

Figure 3.11 *Refraction in a prism.*

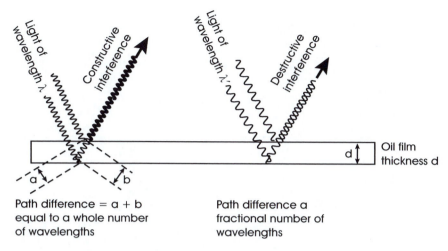

Figure 3.12 *Interference in an oil film.*

different colours. In a prism, the refraction at two non-parallel surfaces prevents the ray being refracted back straight again, leaving the separate colours spread out like a rainbow (see Figure 3.11). Rainbows are caused by this effect, with the sunlight being simultaneously refracted in many raindrops to split out the different colours.

Interference of light causes the rainbow effects in bubbles or when a film of oil is spread over water. When two light waves interact, they may either constructively interfere, when peaks coincide and reinforce one another (termed coherence), or destructively interfere to cancel one another out. Depending on the thickness of the bubble or oil film, reflected light from each side shows constructive interference at a certain wavelength or colour, where exactly a whole number of wavelengths fit across the film. As this thickness varies, the wavelength that 'fits' varies, and so the colours are seen to change through the spectrum.

Diffraction of an EM wave when passing through a small slit causes it to spread out around the edges (see Figure 1.16). A **diffraction grating** consists of a piece of metal or glass ruled with a large number of parallel slits, typically 600 per mm. This makes the spacing a few wavelengths of visible light. When light shines on the diffraction grating, the diffraction patterns from each slit overlap, producing an interference pattern. Where the path difference between two overlapping rays is a whole number of wavelengths, this will produce positive interference, i.e. a bright line. It can be shown that

$$d \sin \theta = m\lambda \tag{3.16}$$

where d is the spacing between slits; θ is the angle of observation; m is a whole integer, and λ is the wavelength, as illustrated in Figure 3.13. With $m = 1$, the principle maximum is seen, which will be brightest and closest to the direct path. If white light is shone on to the grating, a spectrum can be observed, as the angle at

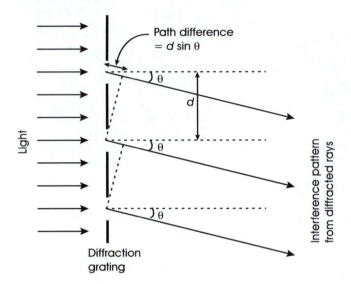

Figure 3.13 *The diffraction grating.*

which interference is observable depends on the wavelength. The diffraction grating can therefore be used to measure wavelength, by measuring the angle to the principle and further maxima, and is a key component in spectrometry.

Light in the atmosphere is subject to scattering, which consists of diffraction around gas molecules, where their size is similar to the wavelengths of light, and reflection from relatively large particles such as dust and water droplets. The amount depends upon wavelength. Blue light is scattered most from gaseous air molecules, which is why the sky is blue. The diffuse light from the sky consists of light scattered by the atmosphere, with more of the blue wavelengths, while looking directly at the sun much of the blue has been scattered out, so it appears yellow or red. At sunset, viewing through the lower part of the atmosphere, much more light is scattered including reflective scattering from clouds and dust, creating the characteristic deep red and orange tints.

Radiation from hot objects: black bodies

All objects emit radiation (as long as they are at a temperature above absolute zero), generally as infra-red (heat). This radiation is all around us all the time, being absorbed and emitted by everything. It is important in looking at heat loss from buildings, or from animals or plants, and in understanding the greenhouse effect.

The amount of heat radiated depends on the temperature, and also on the colour of the object. A black object that absorbs a lot of radiation will also emit a lot; while a white object will reflect radiation, and will not either absorb or emit as much.

Box 3.4

X-ray diffraction

X-ray diffraction (XRD) is the main method used to determine the crystal structure of materials, using a principle similar to the mechanism of the diffraction grating. XRD has a wide range of environmental applications, including determining the structure of proteins and enzymes; identifying minerals in rocks, soils and sediments; detecting cracks and flaws in solid objects and characterising metallic materials such as alloys.

In XRD, a monochromatic beam of X-rays is shone on to a crystal at a certain angle, θ, as shown in Figure 3.14. Some of the X-rays will be reflected from a crystal plane in the lattice, while others continue deeper into the crystal to be reflected from the next plane, or the next. The intensity of the reflected X-ray beam will depend on whether the X-rays constructively interfere. The condition for constructive interference is that the path difference between rays from two subsequent crystal planes is equal to a whole number of the X-ray wavelengths. This gives rise to the Bragg equation:

$$2d \sin\theta = n\lambda \tag{3.17}$$

where d is the distance between crystal planes, θ is the angle of incidence of the X-ray, n is a whole integer and λ is the X-ray wavelength.

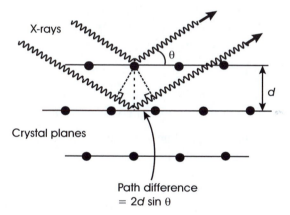

Figure 3.14 *Principle of X-ray diffraction.*

As X-rays will be reflected from a large number of parallel planes simultaneously, at a certain angle constructive interference will occur between all of them, and the reflected X-ray beam will be much brighter, which can be measured with an X-ray detector. X-rays are used rather than visible light as their short wavelength makes it possible to resolve the small distances in interatomic spacing, around 10^{-10} m. The crystal is thus rotated to find angles at which this occurs, and the distance between various parallel planes can be calculated from these angles. This can then be used to determine crystal structure, or it can be used for identification of substances where crystal structure is well known.

So central heating radiators logically should all be painted black to radiate most, and in cold climates everyone should wear white overcoats.

Kirchoff's law states that the amount of radiation emitted from a body at any given wavelength is equal to that absorbed. This has to be true, as otherwise objects at the same temperature as their surroundings could heat up just by being different colours and absorbing more than they emitted without any external heat source. This is clearly impossible as it contravenes the Second Law of Thermodynamics.

A theoretical object that absorbs all radiation hitting it is known as a **black body**, which will also emit the maximum amount. The amount of radiation emitted by a black body at temperature T is given by Stefan's (or the Stefan-Boltzmann) Law:

$$E = \sigma T^4 \tag{3.18}$$

where E is energy emitted per square metre per second, T is absolute temperature and σ is Stefan's constant, equal to 5.7×10^{-8} W m^{-2} K^{-4}.

Most real-life objects absorb less than 100 per cent of all radiation, and so at a certain temperature emit less too. However, energy emitted is still proportional to temperature to the power four, multiplied by a different constant with a lower value. The difference is expressed by multiplying by the **emissivity**, ε, which is the ratio of the actual radiation emitted to that of a black body, and is therefore always less than one. Because of Kirchoff's law, the emissivity also measures the amount absorbed at any given wavelength. For a 'grey' body, the radiation emitted will be:

$$E = \varepsilon\sigma T^4 \tag{3.19}$$

Due to this fourth power radiant heat loss will increase greatly with temperature, e.g. doubling temperature (in Kelvin) would increase heat loss by a factor of 16.

When an object is heated up, the wavelength of radiation given off decreases, and the energy of the radiation increases. For instance, a poker heated in a fire will first be hot but remain black, emitting invisible infra-red, heat radiation. As it gets hotter it will start to glow red, emitting the longest wavelengths of light. At higher temperatures still it will go through orange to white-hot, emitting all wavelengths of light.

At any given temperature it will be emitting radiation over a range of wavelengths (Figure 3.15). The dominant wavelength, at which most radiation is emitted, is given by:

$$\lambda_{max} T = \text{constant} = 0.0029 \tag{3.20}$$

So dominant wavelength is inversely proportional to temperature:

$$\lambda_{max} = \frac{0.0029}{T}$$

Stars vary in colour, from red to blue, according to their temperature. It is possible to estimate the temperature of any hot object – whether a star or the inside of a

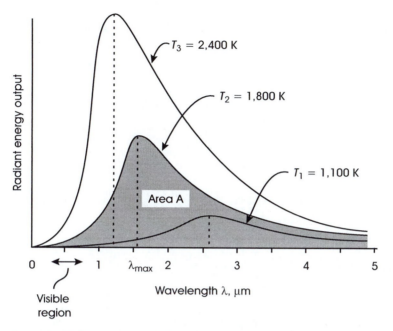

Figure 3.15 _Black body radiation at three different temperatures. Note Area A represents total radiation emitted, given by_ E = εσT⁴ _(shown for T₂)._

furnace – by measuring this dominant wavelength. For instance, radiation from the Sun has a dominant wavelength of around 5.5×10^{-7} m. From this the surface temperature of the Sun can be estimated to be:

$$T = 0.0029/\lambda$$

$$= 5,273 \text{ K}$$

$$= 5,000°\text{C}$$

Conversely the Earth, with a surface temperature of around 288 K radiates at a dominant wavelength of $\lambda = 0.0029/288 = 1 \times 10^{-5}$ m, which falls in the infra-red region of the spectrum. The Earth thus absorbs light from the Sun which heats up the ground, then re-radiates this heat back as infra-red radiation, some of which is absorbed in the atmosphere. At night heat is rapidly radiated from the dark surface of the ground, and takes much longer to be conducted through the soil to cool down deeper levels, or to cool down the atmosphere above the ground. This results in temperatures being most extreme at the soil surface – the coldest place at night, and the hottest during the day (Figure 3.16). It is for this reason that ground frosts commonly occur when air temperatures are a few degrees above zero. The effect is even more marked at altitude, where the atmosphere is thinner and does not absorb and re-radiate the heat emitted as effectively.

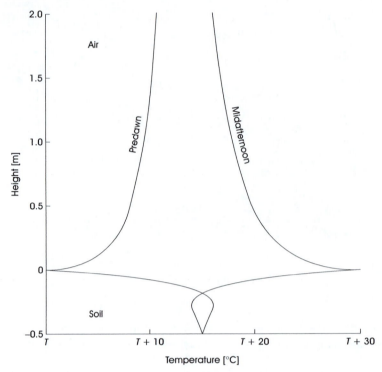

Figure 3.16 *Hypothetical temperature changes close to ground level.*
Source: reproduced by permission from Campbell (1977) p. 10

Atomic absorption and emission spectrometry

The absorption and emission spectra of any element are unique characteristics of that element, which can be measured and used in atomic absorption and emission spectroscopy (AAS and AES). These are powerful tools for environmental analysis, making it possible to quantify tiny amounts of various pollutant molecules or elements in soils, rocks, water and the atmosphere.

Electrons in atoms exist with specific energy levels. If radiation striking the material corresponds to the difference between one of these energy levels and the next, it may be absorbed and excite the electron. As electron energy levels are quantum phenomena this absorption occurs at specific frequencies, such that the energy of the radiation (from Equation (3.15), $E = hf$) is equal to the difference in energy level between the two electron states. These frequencies lie in a wide range from infra-red through visible to UV radiation, and constitute the atomic absorption spectrum.

In a molecule, vibrational and rotational states of the whole molecule also exist, which can be excited at other frequencies, typically in the infra-red or microwave region of the spectrum, giving rise to molecular absorption spectra. X-rays may also be involved, in transitions made by the innermost, most tightly bound electrons in molecules. Molecular spectra are thus rather more complex than atomic spectra.

It follows that when the electron or molecule returns to its former energy state, radiation will be emitted at essentially the same frequency that it was absorbed. This gives rise to the emission spectrum, which may actually be shifted to slightly lower frequencies than the absorption spectrum, due to various quantum effects.

In AAS, the substance to be analysed is exposed to radiation which is then varied over a given frequency range. A diffraction grating (pp. 121–2) is used to spread the radiation out into its component frequencies at different angles. The amount of radiation transmitted through the sample is measured by a detector over the frequency range, to deduce the absorption spectrum. This will be seen as a series of lines at different angles, corresponding to different frequencies. In AES the sample is excited by heating or bombardment with electrons and the resulting emission spectrum analysed.

Spectral lines each have a different intensity and width. Simple atoms with one electron in their outer orbit have simple spectra with a small number of intense, well-defined lines, for instance hydrogen or sodium. For those atoms or molecules where more transitions and interactions are possible, the spectrum becomes more complex, with many lines of differing intensities. The intensity depends chiefly upon the population of electrons in the states being excited. Width of lines depends on a number of factors, including quantum uncertainty (related to Heisenberg's famous principle), collisions between atoms and Doppler shifts dependent on atomic movements. These may be affected by the temperature and in the case of atoms in solution or within some organic medium, interactions with the host will play a part. These factors will help to characterise the spectrum, allowing further information to be gained (Boeker and van Grondelle 1999).

Biological effects of non-ionising radiation

While the high frequency ionising radiation such as X-rays and γ-rays have known harmful effects, discussed on pp. 254–5, lower energy radiation may not be entirely harmless. The energy in non-ionising radiation is low enough not to affect the molecular structure of matter that it interacts with, but it may interact in other ways.

At the lowest energy end, radio waves pass through us all constantly, from radio and TV transmitters, with no apparent ill-effects. Slightly higher frequency microwaves, such as those from mobile phones, penetrate the body and cause slight warming of tissue, as in a microwave oven. Some researchers believe that this has been shown to cause health problems for users of mobile phones especially if used for extended periods of time with the handset held next to the head. It has also been suggested that microwaves have non-thermal effects, being linked to neurological symptoms such as headaches, but there is little statistical evidence to prove these effects (McKinlay 1997; Sienkiewicz 1997).

UV radiation in humans and other animals can cause sunburn, cell damage and skin cancers, including potentially fatal melanoma. Plants and other simpler organisms such as phytoplankton are also susceptible to UV damage, reducing their productivity rates. The UV region of the spectrum is commonly divided into three: UV-A, with wavelengths over 320 nm; UV-B with wavelengths from 290 nm to 320 nm, and UV-C at less than 290 nm. Of these UV-A is least damaging having the lowest energy; while UV-C is normally not present at ground level as it is all absorbed in the atmosphere. While some UV-B is absorbed by the ozone layer (pp. 201–2), the remaining portion is the most damaging biologically, and the region that is increasing as the protective ozone layer is depleted. These wavelengths have sufficient energy to disrupt bonds in DNA and certain other proteins, producing the damaging effects known.

Remote sensing

Many environments that are large or inaccessible at ground level may be monitored from aircraft or satellite by means of remote sensing (RS). Essentially, RS involves using a camera or an electro-optical imaging device to measure radiation from the Earth's surface. This radiation may include reflected sunlight, self-emission due to the temperature of the object viewed, or in the case of active remote sensing, reflected signals from lasers or other radiation sources of specific wavelengths. Many different wavelengths are used in RS, including infra-red, visible and ultra-violet ranges.

While much RS technology has been developed for military purposes, it provides a powerful tool for observing and monitoring many environmental features. Applications are numerous, including studies of ground vegetation and health; forest biomass; climate and temperature monitoring; atmospheric and cloud composition; extent of ice cover in glaciers and polar regions; ocean circulation and sea surface temperature, and ecological data such as algal blooms and plankton levels. Advantages over ground-level studies include measurements of 'invisible' variables such as UV reflected, spatial studies of large-scale areas, and monitoring in otherwise inaccessible areas such as deserts, forests, oceans, icecaps and the atmosphere. To measure atmospheric composition, measurements are taken of self-emission from the atmosphere and compared with the emission spectra of the gases involved to calculate their concentrations. Temporal studies may compare images of a particular area taken over a period of time, revealing long-term changes in climatic or ecological data.

Historically, aircraft and balloons have been used as the chief RS platforms, which are still in use today allowing high resolution over relatively small areas. But space

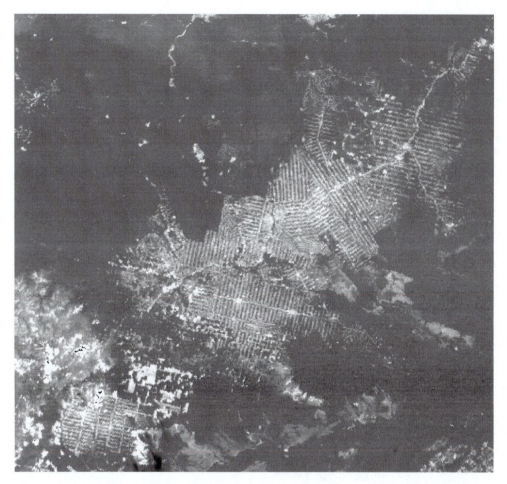

Plate 3.1 *Deforestation of the Brazilian rainforest as shown by thermal RS imaging. This is an 11 μm daytime image, 28 August 1991, covering an area of 512 × 512 km. The cleared areas show up brighter than the surrounding vegetation because they can be up to 4 K warmer during the day.*
Source: photo by courtesy of Rutherford Appleton Laboratory

technology has greatly expanded the possibilities of RS, allowing use of satellite-based systems that can view very large areas. For a description of the orbits used by RS satellites see Box 1.5.

Remote sensing has its limitations. The wavelengths used must correspond to 'windows' in the atmospheric absorption spectrum, in other words wavelengths at which the atmosphere is transparent and allows the radiation to pass from the Earth to the sensor. The principal physical variables being measured are reflectance at various wavelengths and/or self-emission, which indicates temperature, although certain other quantities may be detected. These include polarisation effects (which may be used to identify reflection from water), scattering, orientation or emission of specific wavelengths. To gain useful information about the system observed, these raw physical data must be interpreted to find the variable of interest to the environmental decision maker, which could be vegetative cover, water turbidity, land-use pattern, soil moisture, cloud droplet size or any of a host of others. This process of image interpretation involves many steps and possible means of analysis, according to the application. RS is ultimately limited to features that can be observed remotely. For instance while RS images may identify areas of forests showing decline characteristic of acid precipitation damage, it is not possible to measure the pH of surface waters or the geochemical characteristics of the soil that would determine this damage.

Radiometry in remote sensing

Wavelengths of radiation used in RS depend on the sensor used and the characteristics of the system being observed. An important distinction is between reflectance and thermal imaging. The former relies on radiation reflected from an external source – usually sunlight, but this may be a laser or other source from an active RS system. Thermal imaging relies on self-emission of radiation due to the temperature of the object observed, according to the laws given on pp. 122–5. At a typical ambient temperature of 300 K, from Equation (3.20) the dominant wavelength of emission will be:

$$\lambda = \frac{0.0029}{T}$$

$$= 9.67 \times 10^{-6}\,\text{m}$$

In other words at ambient temperatures, the surface of the Earth radiates infra-red radiation with wavelengths in a range around 10 μm, with hotter objects emitting at shorter wavelengths. Hotter objects will also produce more radiation, according to Equation (3.19), $E = \varepsilon\sigma T^4$.

In the infra-red region, 8–14 μm, self-emission will predominate, while in the UV or visible regions reflection will predominate. Both will be important in the range 3–5 μm, at least by day, making measurements in this range more complicated to analyse.

Early RS systems used ordinary cameras with initially monochrome, later colour film sensitive to three colours of visible light. Modern day systems may use film sensitive to other non-visible areas of the EM spectrum (UV or IR), or electro-optical (EO) detectors that can cover a wide range of wavelengths, from around 0.2 to 20 μm. They have the further advantage over film that digital image information can be radioed back to Earth from the satellite for processing, without the need to physically transfer film.

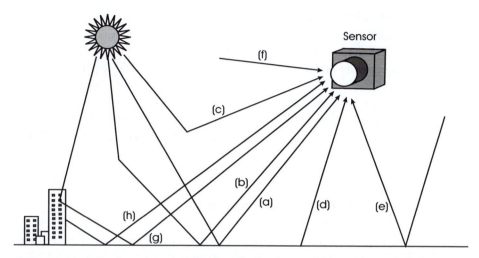

Figure 3.17 *Paths followed by light in remote sensing: (a) direct reflection of solar radiation; (b) solar radiation scattered in the atmosphere followed by reflection; (c) solar radiation scattered in the atmosphere into the sensor; (d) direct self-emission from the target; (e) self-emission from the atmosphere reflected from the target; (f) direct self-emission from the atmosphere; (g) solar radiation reflected from other objects followed by the target; (h) self-emission from other objects reflected from the target.*

Combining the detector with filters allows radiation in several discrete wavelength bands to be measured. For instance, the Landsat Multi-Spectral Scanner (MSS) uses four spectral bands from 0.5 to 1.1 μm wavelengths, corresponding to green, red, short-wave infra-red (SWIR) and long-wave infra-red (LWIR). More recently, imaging spectrometers have been used to vastly increase the number of wavelength bands. The spectrometer is equipped with a diffraction grating to spread the incoming radiation on to an array of detectors. For instance, the AVIRIS (Advanced Visible and Infra-Red Imaging Spectrometer) operated by NASA can measure 224 different spectral channels, covering wavelengths from 0.4 to 2.5 μm with bandwidths of 0.01 μm.

Radiation reaching the sensor will have followed one of several energy paths, as illustrated in Figure 3.17. These include combinations of thermal, reflectance and scattering effects as detailed in each.

The magnitude of each path may be calculated, for a given wavelength band, according to mathematical formulae, taking into account a number of factors, such as the radiation source, how radiation is reflected and spread out, atmospheric effects, and optical effects from the camera system used.

Adding their combined effects produces a large and complex equation that must be solved in order to find the reflectance or thermal self-emission from the object. Fortunately many of these paths will be negligible, depending on the wavelength observed and the application. Paths (a), (b), (c), and (g) originate from solar radiation, while (d), (e), (f) and (h) are thermal and hence of longer wavelengths. For daylight reflectance measurements, the primary path of interest is (a), but adjustments may need to be made for paths (b), (c) and (g). For instance on a hazy day there will be more scattered light, and this will affect blue light more than red; or the presence of a large building, cliff or expanse of water may increase the reflected component (g). Similarly in order to measure target temperature, its self-emission (d) is of primary interest, but allowance must be made for (e), (f) and (h).

Image interpretation and ground-truthing

Having recorded a series of images, a series of data processing steps must be undertaken in order to extract useful information on the parameters of interest, such as vegetation health, etc. The data from the sensor will correspond to brightness levels in the various wavelength bands, either as brightness of a film image or a digital count in an electro-optical imager. These data are converted to temperature or reflectance values, by means of a variety of methods, which may include combinations with ground measurements known as 'ground-truthing'. The image is then analysed and interpreted to identify features and make quantitative measurements, which will again require a range of different methods including ground-truthing, human analysis, mathematical transformations and computer-based operations. For instance quantitative estimates of biomass or primary productivity of phytoplankton can be made from data on the absorption in specific wavelengths, corresponding to the wavelengths used in photosynthesis.

For thermal RS, principles applied to find temperatures include correlation with ground-level measurements of temperature, and use of theoretical models, amongst others. In making reflectance measurements, there are additional methods specific to reflectance, such as the use of ground-truthing control panels, consisting of flat black, grey and white boards of known reflectance, large enough to be seen on the image, to which the output can be calibrated.

Various computer techniques are available for image processing, in the interpretation stage. Electro-optical imagers produce digital image data, often far too complex to be comprehensible directly as an image. A single image would be divided into pixels – small grid squares, each representing say a 10×10 m patch of ground, with a brightness value. As there may be up to 224 wavelength bands, each pixel may represent up to 224 values. The amount of data needed can be very large. For instance a single Landsat Thematic Mapper image of $6,000 \times 6,000$ pixels with seven spectral bands takes up 250 MB of computer memory, for ground coverage of 185×185 km, divided into 30 m pixels. To reduce processing time, the analysis may concentrate on specific areas of interest or limit the spectral bands considered.

Algorithms can mathematically enhance images by intensifying edges and contrast, smoothing to filter out noise, or identification and highlighting of specific patterns and features such as water or a certain forest type. Differences can be quite subtle, for instance a difference in reflectance in the infra-red band may distinguish between crop types. A computer can detect this readily while it would be invisible to the naked eye.

Where a large number of spectral bands are available, the user may need either to select specific bands of interest, or average the data into a smaller number of wider bands. This both reduces processing needed and may reduce errors in 'noise' in the signal, without losing a great deal of useful information. Accuracy may be limited by trade-offs between different factors, including cost, computing power or human analyst time available, and physical limitations. A key physical trade-off is that between the spatial, spectral and radiometric aspects of the data. Given there is a limited amount of radiation available to measure, if it is divided into a large number of spectral (wavelength) bands and into many high resolution, small sized pixels per square kilometre, there will be less accuracy possible in the brightness level measured from each.

Geographical information systems (GIS) are ideal for analysis and display of RS images. A GIS system can produce digital maps of the surveyed area, overlying RS data on top of that from other sources, such as rock or soil type, topography, settlement patterns or climate. The system may then be used to identify areas of interest that combine these different types of information – for instance heather moorland over 1,500 m in height; coniferous forests on acid rocks with over 2,000 mm annual rainfall, or oceans with a temperature above 15°C and more than 200 km from land. Mathematical functions can be applied, to calculate areas in the specified categories, and look at the possible implications of environmental policy decisions.

Summary

- Heat is the energy of molecules vibrating, governed by the laws of thermodynamics.

- The heat transfer processes of convection, conduction and radiation each have physically distinct mechanisms. These determine heat losses from buildings, and are important in many natural systems, including heat balance in animals and climatology.

- The internal combustion engine is the commonest form of heat engine, converting from heat to useful mechanical work. The Carnot cycle describes a configuration of pressure, volume, energy and entropy in such a system, allowing thermodynamic efficiency to be estimated.

- Electromagnetic radiation includes X-rays, light, radiant heat and radio waves, which all have some properties in common.

- Remote sensing uses satellite-based instruments to detect reflected and emitted radiation from the Earth's surface, characterising environmental systems and allowing study of many features that can be deduced from colour and temperature.

Questions

1   A child builds a big snowman which contains 50 kg of snow. How much energy would it take to melt it, from an initial temperature of −2°C? Why doesn't it melt until long after the rest of the snow has gone?

2   (a) How is heat lost by a cup of coffee?

    (b) I have just poured myself a coffee but not yet added the milk, when there is a knock at the door and I have to go and see someone. I'll be gone a few minutes – should I add the milk before I go or after I get back in order to have the coffee at the maximum temperature when I drink it?

3   Which of the following are types of electromagnetic radiation?

    (a) light;
    (b) sound;
    (c) X-rays;
    (d) radiant heat;
    (e) TV signals;
    (f) microwaves;
    (g) waves on the sea.

4   The surface of a central heating radiator is at 47°C (320 K) and it is found to radiate 500 W of heat. If its temperature is increased to 57°C, what will the radiant heat output increase to?

5   A glass of water containing 0.25 l is filled up to the brim from tap water at 10°C. It is then left to warm up to room temperature of 20°C.

    (a) How much energy will be used in warming up the glass of water?

    (b) When it expands due to this warming, how much will spill?

    (c) If the glass had initially contained a mixture of ice and water at close to freezing point, and was put in a sauna at 80°C, could you use the same method to answer (a) and (b)?

6   How is heat lost in hot climates by animals? How is heat retained in cold climates? Give examples of animals that display hot and cold climate adaptation, in features or habits.

7   Which waves carry most energy – UV light, infra-red, or radio waves?

8   Two cans of drink are left out in the sun on a hot day. One is in a dark-coloured can, while the other can is silver and white in colour.

    (a) After an hour, will the two drinks be at different temperatures, and why?

    (b) Might either drink get hotter than the surrounding air?

    (c) What stops them getting any hotter than they do?

9   A solar water heater absorbs sunlight and converts it to hot water at 40 per cent efficiency. The area of the solar panel is 4 m^2 and the Sun's energy is falling at 800 W m^{-2}. How long does it take to heat 50 l of water for a shower from 15°C to 60°C?

10   A greenhouse has a total area of glass of 30 m², single glazed.

(a)   Neglecting losses through the aluminium frames, if the greenhouse is at 20°C inside and it is 10°C outside, how much energy is lost by conduction? How else is heat lost? Where does this energy come from?

(b)   What power of electric heater would be needed at night, to maintain a temperature of 20°C, if it is still 10°C outside (neglecting non-conductive losses)?

(c)   The greenhouse contains a black plastic, 150 l capacity water butt. How much energy will the water butt lose to the greenhouse if the water cools by 1°C? Will the water butt make any difference to the temperature inside the greenhouse, during the day or at night (without the electric heater)?

Answers to numerical parts

1   16.9 MJ

4   565 W

5   (a)   10.5 kJ   (b) 0.375 ml

9   1 hr 8 mins

10   (a)   1.68 kW   (b) 1.68 kW   (c) 630 kJ

Further reading

Principles of Environmental Physics (2nd edn). J. L. Monteith and M. H. Unsworth. Edward Arnold, London, 1990. Contains chapters on heat flow in plants, animals and soils, plus useful material on radiation.

Energy Management. P. W. O'Callaghan. McGraw Hill, New York, 1992. Covers energy conservation and management in buildings.

Environmental Physics (2nd edn). E. Boeker and R. van Grondelle. John Wiley & Sons, Chichester, 1999. Contains a very detailed section on spectroscopy and also other topics including thermodynamics, all at a more advanced level than given here.

Remote Sensing – The Image Chain Approach. J. R. Schott. Oxford University Press, Oxford, 1997. Excellent guide to remote sensing theory and application.

4 Solids, liquids and gases

It is the physical characteristics of matter contained in an environmental system that determine many of the key processes and flows within that system. The following key concepts are covered in this chapter:

- The physical states of matter – solids, liquids, gases and plasma – are determined by intermolecular forces, dependent upon temperature and pressure
- To change state involves exchange of energy and may also need 'seeding'
- Strength and elasticity are important characteristics of solids, which may be applied to rocks
- In a gas, pressure, temperature and volume are closely related, giving rise to the 'gas laws' which are important in describing energy flows in the atmosphere
- Fluid dynamics, the study of movements of fluids, is applicable to a wide range of environmental systems

In solids, including rocks and soils, factors such as strength and elasticity are important in geotechnical and geomorphological contexts. The processes governing heat flows and pressure changes in gases are subject to specific laws, crucial to an understanding of atmospheric processes. Movements of pollutants in water or air, aspects of climatology and the flight of birds can all be described by the complex field of fluid dynamics. Groundwater presents a specific case for the environmental scientist, having a key role to play in transport of contaminants and in provision of clean water within natural ecosystems and for our use.

States of matter

Matter can exist in four different states: solids, liquids, gases and plasma. The differences between these are due to changes in the way the molecules interact, which depends on the balance between each molecule's thermal energy and the electrical forces between the molecules – that is, a balance between kinetic and potential energy.

In a **solid**, the molecules have relatively little thermal energy and are bound together into a fairly rigid structure by the forces between them. The molecules may vibrate, but not move within the solid. Some solids are plastic, allowing deformation or bending without breaking, while others are rigid and brittle, with a much more fixed structure. As the temperature increases, the molecules vibrate more strongly and start to break free of this structure, forming a **liquid** which can

(a)

(b)

(c)

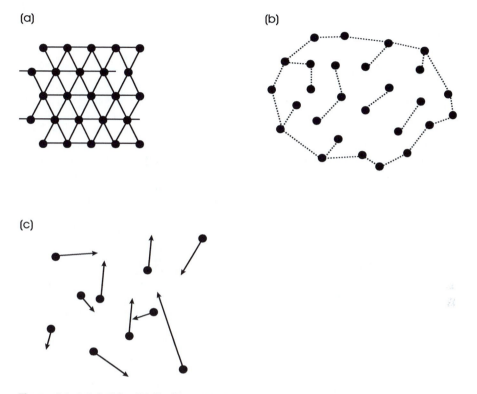

Figure 4.1 *(a) Solids, (b) liquids and (c) gases.*

flow, but which holds itself together and has a surface due to attraction between the molecules. At higher temperatures, this too breaks down to make a **gas**, and individual molecules can move around freely with little interaction between them. A gas thus has much lower density and no coherence or fixed surface. Liquids and gases are classed together as fluids, which share certain properties such as diffusion and flow.

These are the three states of matter commonly encountered. But if a gas is heated to a temperature of several million degrees, the thermal energy will be sufficient to completely ionise it, that is to free all the electrons from the nucleus. The gas now becomes a **plasma**, with very different properties, as the nuclei are no longer protected by their electrons and are more likely to collide and take part in nuclear reactions. Plasma exists in particle accelerators used to study elementary physical particles, in the Sun where nuclear reactions take place, and also at much lower densities in the upper atmosphere, where very low pressure gases are ionised by solar radiation. The aurora borealis or 'Northern Lights' is a visible plasma phenomenon, which occurs due to interaction of particles excited by the solar wind and the Earth's magnetic field.

Box 4.1

The structure of the Earth

The Earth is not completely solid, but consists of a number of layers with different physical and chemical properties (as shown in Figure 4.2). Chemically, the Earth is differentiated into three zones: the core, which is mostly iron, (thought to be in the form of iron–nickel alloy in the inner core and iron–silicon alloy in the outer core); the mantle, containing ferromagnesian silicates and oxides; and the crust. The continental crust is a thin surface layer, between 20–40 km in thickness (greatly exaggerated in Figure 4.2), whose composition is diverse, containing many lighter elements such as aluminium, potassium and sodium. Oceanic crust is thinner, averaging about 8 km depth and has a different composition. This structure reflects conditions when the Earth formed from a liquid ball, when lighter elements could float to the surface.

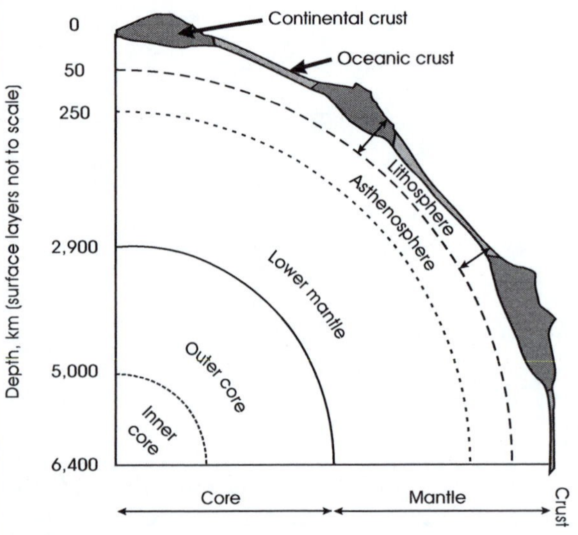

Physically, the uppermost solid layer is the lithosphere, which comprises the crust and the upper 50 km or so of the mantle. This region is rigid, and makes up the plates that move around the surface, giving rise to plate tectonics. The plates 'float' on the asthenosphere, a part of the mantle where temperatures are high enough for rocks to be plastic or partially molten. Below about 250 km, as pressure increases, the mantle becomes solid once more. The outer core is liquid, as the iron alloy has a lower melting point than rock, while the immense pressure in the inner core keeps it solid. Pressures in the core are between 1 million and 3.5 million times atmospheric pressure while the core temperature is between 5,000–6,600°C, both temperature and pressure increasing with depth (Montgomery 1995). It is thought to be currents in the liquid iron of the outer core that give rise to the Earth's magnetic field, arising from a combination of convection and the rotation of the Earth.

Figure 4.2 The structure of the Earth.
Source: adapted from Montgomery (1995)

The average density of the Earth is about 5,500 kg m^{-3}, a figure that is about twice the density of crustal rocks. It is evident that the lower mantle and core are compressed to a density higher than that found at the surface.

Solutions and other mixtures

In most real world situations materials do not exist in their pure state but as various mixtures and solutions.

In a **solution** the molecules or ions of one material (which could be solid, liquid or gas) are evenly spread out through a liquid, as opposed to a **suspension** where small particles of a solid are mixed in a liquid. A suspension can be separated by filtering or centrifugal methods, a solution cannot. The rate at which the suspended solids settle depends principally on the particle size. In a natural setting such as an estuary, this leads to sorting of mud and sand grains by size into layers in the sediments, which will be retained in sedimentary rocks.

An **emulsion**, as in vinaigrette dressing, is a suspended mixture of insoluble liquids such as oil and water (or vinegar), brought about under specific chemical conditions together with physical mixing to break the oil up into tiny droplets. Emulsions will settle out over time, or in a centrifuge.

Some organic materials such as proteins may have more specific properties when mixed with water, forming colloidal suspensions or gels. Water is chemically held by the protein molecules, forming a mixture that may maintain some of the solid properties such as strength, shape or stickiness, e.g. the shape of jelly in a mould. These properties are used by many biological organisms, from mucous membranes protecting our lungs, to slug and snail trails and microbial protective slime.

A liquid can also evaporate to mix with a gas, like water vapour in the atmosphere, which resembles a solution in some ways. Water is present in the atmosphere in several forms – as a gas, water vapour; as a liquid in the form of clouds, mist and fog; and as a solid as ice, snow or hail. Where liquid water is held in tiny droplets, too small to fall as rain, this is known as an **aerosol**. Aerosols are important, as the liquid water can dissolve gases and other substances, sustain chemical reactions, and transport living viruses and spores.

The maximum amount of a material that can be dissolved in another is known as the **saturation point**. This term can be used for a material dissolved in a liquid, or for a liquid evaporating to mix with a gas. Saturation point increases with temperature, in the same way as you can dissolve more sugar in cup of coffee when it is hot. For gases dissolved in liquids however it decreases with temperature – warm water can dissolve less oxygen than cold, just as warm lemonade goes flat more quickly than cold.

The freezing point and boiling point of a liquid solution will differ from that of the pure liquid. For instance salt water freezes at a lower temperature, down to $-22°C$ depending on the amount of salt in solution, and boils at a higher temperature. But when the solution evaporates or freezes, the pure solvent will separate out. Thus sea ice contains pure fresh water and the remaining sea water underneath is saltier – important in ocean convection (p. 199). The exception is a mixture of two substances close in boiling point, such as water and alcohol – in this case the alcohol starts to boil at about 80°C, and the proportion of water to alcohol in the steam will increase until the temperature reaches 100°C.

Changes of state

When a solid melts, a liquid evaporates or a gas condenses it has changed state. These changes are not just due to a change in temperature. To melt or evaporate, matter needs extra energy to break molecular bonds, which is the latent heat (see Chapter 3). But in addition other factors affect changes of state.

Many substances start to evaporate at temperatures well below their boiling point, even solids, which is how we smell things. A substance that is **volatile** evaporates readily at room temperature, which may be well below its boiling point, indicating a low latent heat of evaporation. The air contains water vapour at room tempera-ture. The amount of liquid water evaporating depends how close the concentration of water vapour in the air is to the saturation point, the maximum amount the air can hold. The saturation point increases with temperature, and so the evaporation rate is also highly dependent on temperature, increasing more rapidly above about 40°C. This is relevant to climatic systems, as increases in temperature brought about by climate change will increase evaporation rates quite significantly, with impacts on rainfall, cloud cover, soil moisture content, energy transfer as latent heat and many other considerations. Water cannot exist as a liquid above 100°C at normal atmospheric pressure so at this temperature any more heat applied will evaporate the remaining liquid. At low pressures (such as at altitude), the boiling point of water is reduced – which is why mountaineers have to boil their pasta for longer.

At low temperatures, below freezing point, water vapour in the air will readily condense and freeze (forming frost). In cold climates the air is thus very dry, and dehydration can become a problem for people and plants. Evergreen trees such as conifers need their hard waxy needles to prevent water loss in the winter.

High pressures can maintain a gas in a liquid form at room temperature, as in bottled propane or LPG (liquefied petroleum gas). Similarly in the Earth's mantle at depths of a thousand kilometres or so, high pressures keep the mantle rocks solid, despite their very high temperatures that would result in melting at the surface (see Box 4.1).

The melting point of ice is also affected by pressure, but in this case it decreases. Ice is unusual in that it expands when it freezes, thus the tendency to compress ice from pressure will tend to melt the ice. In fact the melting point of ice decreases by about 7.1×10^{-8} °C Pa^{-1}, where the Pascal (Pa) is the unit of pressure (see p. 152). Ice-skaters rely on this – the pressure from the skate results in a thin film of water between skate and ice, which allows the skater to slide and glide easily. Very cold

ice would not have this film of water and so would not be as slippery. Under a glacier, this process is of considerable importance. The weight of the glacier produces high pressures that are sufficient
to melt a thin layer on its lower surface, allowing it to flow and move over the ground. Friction is also important, as the heat produced also aids melting of the ice.

Cloud seeding and condensation nuclei

A gas will condense much more readily if it attaches to small particles known as **condensation nuclei** around which it can form droplets. This process is known as **seeding**. Similarly, a liquid will solidify more readily if seeded, so ice will form around existing crystals, giving rise to snowflakes and 'jack frost' patterns on frozen windows.

In fact, without some form of condensation nuclei a gas cannot condense, nor a liquid freeze. If a mixture of vapour and gas (e.g. water vapour and air) is cooled to below its saturation point and there are no condensation nuclei present, it will become **supersaturated**, as will a solution in a liquid cooled below its saturation point. Introduction of an impurity will then result in condensation of the gas or crystallisation of the solid. Similarly, a pure liquid may be cooled below its freezing point, in which case it will become **supercooled**. If a crystal of the solid is added it will allow freezing/crystallisation of the solid, resulting in an increase in temperature as latent heat is released. Liquids may become superheated, for instance a cup of coffee reheated in a microwave oven may suddenly and explosively produce large amounts of vapour when it is stirred.

Water vapour in the atmosphere will condense under the right conditions to form clouds or mist, but clean air can readily become supersaturated. Dust and pollution in the atmosphere can provide the seeding necessary for condensation. A visible illustration is the vapour trail from an aeroplane (Plate 4.1). On a clear day these vapour trails can make huge and lasting tracks across the sky. The trail is not just the exhaust gases from the plane – the burnt fuel in the exhaust provides particulates around which water in the atmosphere can condense. Smog forms in the same way – smoke and dust from urban pollution provide condensation nuclei that then creates fog, in the form of Los Angeles' constant haze or the infamous pea-soupers of 1950s London.

An interesting paradox is that clouds seeded in this way from pollution particles do not produce as much rain as clouds in cleaner air. In polluted air, the large number of condensation nuclei results in a large number of small water droplets, that do not grow large enough to fall as rain. Clean clouds have smaller numbers of larger droplets, that are more likely to fall as rain, typically twice the size of those in polluted clouds (Rosenfeld 2000).

Plate 4.1 *Vapour trails from aircraft.*
Source: photo taken by the author in Cornwall, November 2000

Box 4.2

Smog in a bottle

You can create your very own pollution blackspot, using just the following:

- a large bottle, like a storage jar or sweet jar;
- a rubber glove; and
- a box of matches.

Put a little water in the bottom of the bottle, and then fit the rubber glove tightly over the top. If you put your hand in the glove and then pull it up and inside out, this will reduce the pressure and hence the temperature inside the bottle, which should make water vapour condense. But you will not see anything occurring, because in order to condense the water vapour needs some form of condensation nuclei.

Now remove the glove, and drop a few lighted matches in to create some smoke. Replace the glove, and try again to pull it up and inside out. Because of the condensation nuclei from the matches' smoke, you will now see the water condenses into fog – and by releasing the glove you can make the fog disappear again.

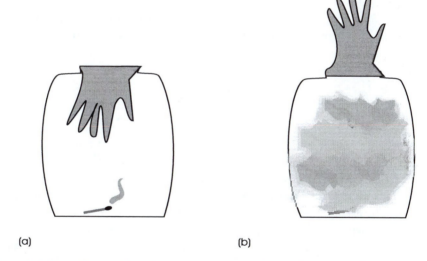

(a) (b)

Figure 4.3 *Smog in a bottle: (a) match produces condensation nuclei; (b) pulling out the glove reduces pressure and temperature, creating smog.*

Molecular diffusion

If a gaseous pollutant is released into still air at a certain point it will gradually diffuse out into the surrounding atmosphere, as a result of the random, thermal motion of molecules. Likewise a liquid pollutant will diffuse into a body of water, and carbon dioxide from the atmosphere gradually dissolves in the oceans and diffuses downwards. Molecular diffusion is a process quite distinct from mixing by wind or water currents, or from turbulent diffusion, where fluids are mixed and dispersed by turbulence. It occurs principally in gases and liquids, and can be a contributory mechanism in contaminant transport in the atmosphere, surface waters and groundwater. In most conditions however turbulent mixing and flow due to pressure and convection (i.e. wind and currents) will be more important in transporting the bulk of the pollutant, although diffusion may cause the plume of pollutant to spread out over time. Diffusion is important at a cellular level in the transport of dissolved gases within an organism, such as from the surfaces in the lung to the bloodstream, or within a leaf. It is also important underground, in groundwater and through pore spaces in rocks for gaseous pollutants such as methane or radon and long-lived, soluble radioactive wastes.

Molecular diffusion is slow. Unlike convective mixing or physical stirring, the movement of each molecule is random and separate, travelling in what is known as a **random walk**. Imagine a drunk staggering around – two steps forward, one step back – and that is what gas molecules are doing as they continually change direction, colliding into one another. Like the drunk it takes a long time to get any great distance in one direction.

The rate of diffusion depends on the **concentration gradient**, that is the rate of change of concentration with distance, together with temperature, pressure and the molecular weight of the substance concerned. Fick's law states that the diffusion rate is proportional to the concentration gradient:

$$Q = k_D \, \Delta C / \Delta z \qquad (4.1)$$

$$\approx k_D \, (C_2 - C_1)/(z_2 - z_1) \qquad (4.2)$$

where Q is the mass flux of material diffusing per unit area (in kg m^{-2} s^{-1}), $\Delta C/\Delta z$ is the concentration gradient in the direction z, and k_D is the diffusion constant or diffusivity (dependent on temperature, pressure, material and so on), which has the units m^2 s^{-1}. Over a short distance and at low concentrations, the concentration gradient can be approximated by the ratio of the differences in concentration C to distance z, as shown in Equation (4.2). The rate is fastest from a concentrated source of pollutant into surrounding clean air or water. As the substance diffuses outwards, the concentration gradient will decline, and so the rate at which it diffuses slows further. This leads to concentrations showing an exponential decline as distance away from the source increases. Note that unlike for fluid flow, there is no need for continuity of flow – the flux declines as the distance from the source increases. Taking one dimension only for simplicity, concentration can be expressed as a function of the distance from the source, taking the form:

$$C(z) = C_0 \exp[-\tfrac{1}{2}(z/\sigma)^2] \qquad (4.3)$$

where C_0 is concentration at source, z is the distance from that source and σ is a measure of the typical distance diffused after a certain time. Over time the amount diffused will increase and the distance will increase.

Box 4.3

Diffusion of methane from landfill

Methane produced in landfill sites can be a serious environmental hazard, as it can diffuse through neighbouring rocks and accumulate under houses or other nearby buildings. If concentration reaches a critical level, it can explode, which has caused serious building damage on occasion. For this reason, methane must be vented from landfills, and the risk of diffusion to neighbouring buildings assessed. Predictive calculations are useful in planning development, as if migration occurs remedial action is expensive.

For instance for the case of houses located 50 m from a landfill site, where methane concentrations within the site are measured at 4.1 per cent, the flux of methane diffusing to the houses is required. The site is built on sandstone. Diffusivity of methane in sandstone is 1×10^{-6} m^2 s^{-1} under normal temperatures and pressures (O'Riordan and Milloy 1995).

Assuming typical background methane concentration in air to be 1.7 ppm, Fick's law from Equation (4.2) can be used to estimate the flow (here expressed in volume terms):

$$Q = 1 \times 10^{-6}(0.041 - 1.7 \times 10^{-6})/50$$
$$= 8.2 \times 10^{-10} \, \text{m}^3 \, \text{m}^{-2} \, \text{s}^{-1}$$

An unsealed basement will receive this flux according to the exposed area, and must be ventilated accordingly to prevent build up. A minimum building ventilation rate is one air change per hour. For instance over an area of 100 m^2, over an hour (3,600 s) the flux would contribute 2.95×10^{-4} m^3 of methane. For a 2 m deep space giving volume of 200 m^3, this would result in an increase in concentration of 1.48 ppm. This compares to 1.7 ppm in normal background air.

This calculation is an approximation, as the flux will actually decrease as concentrations fall with distance away from the site. Also many other factors may be important in increasing flow by means other than molecular diffusion, such as fissures in the rock, pressure from the build-up of gas in the site, or the presence of water.

Molecular diffusion is analogous to conduction of heat, and the diffusivity in Fick's law (Equation (4.1)) is the equivalent of thermal conductivity discussed on pp. 98–9. As the physical transfer mechanism is the same, of movement via collisions between molecules, a material with a high diffusivity will also have high thermal conductivity, and the relationship with temperature will be similar for both constants.

Properties of solids

Solids are characterised by their shape, strength and resistance to external stresses. Their properties determine the materials used in applications from construction to everyday objects, and also the organic materials such as wood and bone that have evolved in the natural world.

Elasticity, stress and strain

Stress on a material means how powerful the forces are trying to deform it, defined as the force per unit area. **Strain** is deformation caused by a stress in an object, as a proportion of its length, e.g. the extent to which an elastic band can be stretched. For certain materials over a range of stresses, strain is proportional to stress. For a wire being stretched, the relationship is described by Young's Modulus, E, as in Figure 4.4(a):

E = stress/strain

strain = extension/length

$\quad = e/l$

stress = force/area

$\quad = f/A$

so Young's modulus

$$E = \frac{(f/A)}{(e/l)} \tag{4.4}$$

Knowing Young's modulus allows the calculation of strain due to a given stress, and indicates the elasticity of the solid. Similarly we may define a bulk modulus, K = bulk stress/bulk strain (Figure 4.4(b)), and a shear modulus, μ = shear stress/ shear strain (Figure 4.4(c)). Of these the bulk modulus will generally be lowest, as any material will be much more resistant to compressive stresses. When a solid block is subjected to a force it will flatten and broaden as in Figure 4.4(d). Poisson's ratio, denoted σ, is defined as the ratio of this broadening to compression, and is an important factor in determining the strength of rocks for large civil engineering structures.

Young's modulus and the bulk compression modulus may also be defined for a liquid or a gas, however in these cases the shear modulus will be zero as fluids have no resistance to the shape changes caused by shear stress.

The various moduli are not independent, and in fact the shear and bulk moduli can both be written in terms of Young's modulus E and Poisson's ratio σ.

Shear modulus:

$$\mu = \tfrac{1}{2} E/(1 + \sigma) \tag{4.5}$$

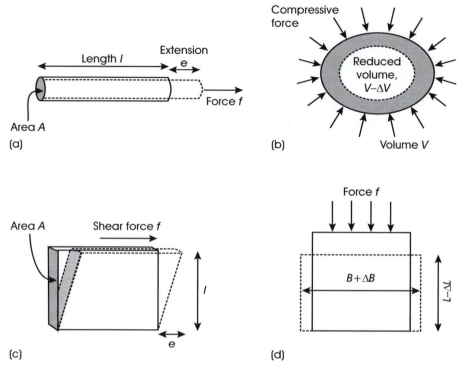

Figure 4.4 *Elastic deformation: (a) linear extension; (b) volume compression; (c) shear strain; (d) Poisson's ratio, $\sigma = (\Delta B/B)/(\Delta L/L)$.*
Source: adapted from Sharma (1997) p. 116

Bulk modulus:

$$K = \frac{E(1 - \sigma)}{3(1 + \sigma)(1 - 2\sigma)} \tag{4.6}$$

These elastic moduli are only applicable over a given range known as **elastic deformation**, when the material will return to shape after the stress is removed. The material is storing energy which is released as it returns to shape – as in a spring or a crossbow. Molecular bonds are being stretched but not broken. For a tensile stress on a wire, elastic strain energy is given by:

$$\text{energy} = \text{force} \times \text{distance (extension)}$$

$$= (\text{average stress} \times \text{area}) \times (\text{strain} \times \text{length})$$

$$= \tfrac{1}{2} \times \text{stress} \times \text{strain} \times \text{volume}$$

Past the elastic limit **plastic deformation** occurs, when the material will not return to shape. Molecular bonds break, or planes slide against each other. Energy is released as heat. If the stress is continued, the material will reach its breaking point.

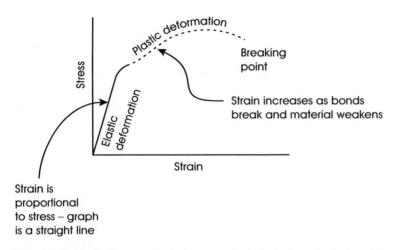

Figure 4.5 *Illustrative graph of stress against strain for a typical metal wire.*

Strength

The strength of a material depends upon the shape of the curve in Figure 4.5. A brittle material such as china will display very little elastic or plastic behaviour but break suddenly if sufficient force is applied. Soft materials such as Plasticine are plastic over a large range, while rubber is elastic over a large range. Elasticity is important in many applications, allowing the material to bend without breaking, as for instance a tall tree bends in strong wind, absorbing some of the wind's force without being blown over.

These properties will vary with temperature, with many materials exhibiting more elastic or plastic behaviour at higher temperatures. Strength is also very dependent upon purity, as a small number of impurity atoms can greatly change the intermolecular properties. This can be used to advantage in producing materials such as carbonised steel where a small proportion of carbon atoms amongst the iron greatly increases tensile strength. It is important to note that recycled materials (such as metals) often require purification to retain the properties of the virgin material. The molecular structure may also vary after time or repeated movement or heating and cooling, causing failure due to metal fatigue and related syndromes. Long-term exposure to light or UV can affect strength in some biological materials such as rubber, while ionising radiation can affect physical properties by changing chemical and nuclear structure, a consideration of importance in the nuclear industry.

Stress and strain in rocks

Rocks exhibit both elastic and plastic behaviour under appropriate conditions, which contribute to the formations underlying natural geomorphological features. The surface crust of the Earth undergoes continual, slow movements as the continental plates 'drift' across the Earth's surface, and new crust is formed in the deep ocean rifts. When plates collide, they exert great pressure on one another, as one slides under the next and major features such as oceanic trenches and mountain ranges are formed.

A rock's plasticity may be much higher at higher temperatures found underground, particularly close to active volcanic regions around crustal plate margins. Over long timescales of continued pressure, solid rocks can bend and be uplifted in this way, creating the distorted and sloping strata found in many rock formations, illustrated in Figure 4.6. If the forces exceed the plastic limit however the rock will break. This creates fissures and cracks, and the sudden movements are experienced as earthquakes.

Rocks also exhibit elastic behaviour. A rock may distort under the action of seismic forces, storing energy as elastic strain energy in the rocks. When the force finally becomes greater than its breaking strain it will suddenly break and spring back, releasing a large amount of energy as an earthquake. This process produces the discontinuities in strata found at the fissure or fault line. Another example is glacial rebound in certain areas such as Scandinavia, Canada and Scotland since the last ice age. This process is due to isostasy or isostatic compensation (see Box 1.2), as the weight of the ice on the land distorted the lithosphere downwards as if it were floating. Since the ice melted around ten thousand years ago, the rocks have been gradually springing back into shape, resulting in an increase in the height of the land and leaving obvious coastal features such as raised beaches, high and dry above sea level. Uplift rates in parts of Scandinavia are as much as 10 mm year^{-1}.

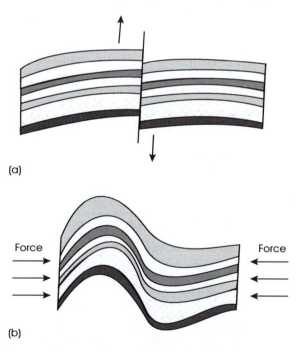

(a)

(b)

Force

Force

Figure 4.6 *Illustrative rock strata formations: (a) elastic rebound along a fault; (b) plastic behaviour in folding of sedimentary rocks due to tectonic compression.*

The strength of rocks is of importance in large civil engineering projects such as large buildings and in dams, where the weight of the structure and the water must be supported by underlying rocks without causing failure of the dam. The strength of rocks is assessed by measuring Young's modulus and Poisson's ratio, together with surveys to find any major faults.

Seismic waves and seismology

Seismic waves, discussed on pp. 33–4, form due to the application of stress from a sudden movement in crustal rock. Earthquakes occur when rocks that have been under stress break and spring back, producing waves that propagate in different directions. The longitudinal P-waves are produced in the direction of the movement, as a result of the compressional strain, while S-waves are propagated perpendicular to it, as a result of shear strain.

The elastic properties of the rocks involved determine the propagation of these waves, and hence their velocities can be expressed in terms of the rock's elastic moduli and its density. Using Equations (4.5) and (4.6) they can thus be represented in terms of Young's modulus E and Poisson's ratio σ (Richter 1958).

Velocity of S-wave:

$$V_s = \sqrt{\frac{\mu}{\rho}} = \sqrt{\frac{E}{2(1 + \sigma)\rho}} \tag{4.7}$$

Velocity of P-wave:

$$V_P = \sqrt{\frac{K + 4\mu/3}{\rho}} = \sqrt{\frac{(1 - \sigma)E}{(1 + \sigma)(1 - 2\sigma)\rho}} \tag{4.8}$$

where μ is shear modulus, K is bulk modulus and ρ is density.

From this it can be seen that if μ is zero (as in a fluid), the velocity of S-waves will be zero, i.e. they cannot propagate; also that the velocity of P-waves is always greater than the velocity of S-waves (given that K and μ are positive) as:

$$V_P^2 = \frac{K + 4\mu/3}{\rho} = \frac{K + \mu/3}{\rho} + \frac{\mu}{\rho}$$

$$V_P^2 = \frac{K + \mu/3}{\rho} + V_S^2 \tag{4.9}$$

Seismic wave velocities can be measured and used to deduce rock properties, either from natural seismic waves from earthquakes, or from waves produced by explosions. This latter method, known as explosion seismology, is most often used in an environmental and geotechnical context to study crustal features such as extent of different rock types, faults, fissuring, the level of the water table, hydrocarbon exploration and site investigations for construction of building foundations or underground waste sites. An important application is the identification of suitable sites for a deep repository for nuclear wastes, as features such as rock type, fissuring, faults, porosity and water content can all be examined at depths up to several kilometres, at far lower cost than deep drilling.

The wave properties relevant to seismology are reflection and refraction. When an S-wave or a P-wave meets a boundary (which may be an interface between different rock types, a fault line or the water table), it will be both reflected and refracted. The interaction between the wave and the boundary will produce both S- and P-wave types from the incident wave. A single incident wave can thus produce four new waves at a boundary: reflected S- and P-waves and refracted S- and P-waves, as in Figure 4.7(a). The angles are determined by an adapted form of Snell's law (p. 31), stating that the ratio of the sine of the angle to the velocity of each wave is constant. It follows that for a certain value of the angle of incidence the angle of refraction will be 90°, such that $\sin \theta_c = V_1/V_2$, where θ_c is the critical angle and V_1 and V_2 are the wave velocities in each rock type. At this angle the refracted wave will travel along the boundary, and as it does so it will produce secondary waves that re-propagate back up to the surface as in Figure 4.7(b).

These properties are used in reflection and refraction surveying. In the former, the seismologist measures the time taken for seismic waves to be vertically reflected back up from some underground boundary. Reflection surveying can be used at depths from tens of metres up to tens of kilometres, allowing a wide range of applications. In refraction surveying, the source and detector are separated by some distance and the waves travel predominantly horizontally along the surface layers, allowing its use over large areas for reconnaissance surveying. The technique relies upon measuring arrival times of secondary waves produced from a wave refracted at the critical angle.

Seismic waves will be attenuated as their energy is lost through friction, and as they spread out away from the source. Lower frequency waves are found to travel

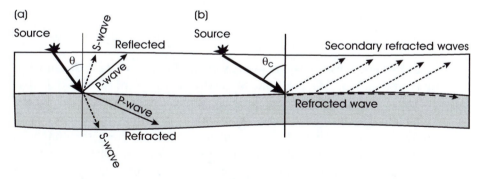

Figure 4.7 *Reflection and refraction, and the critical angle: (a) waves produced at a boundary; (b) refraction at the critical angle.*

further, while the high frequency waves lose their energy more quickly. However high frequency waves allow for better resolution in mapping of features.

Rock types display characteristic S- and P-wave velocities that can be used in identifying them. P-wave velocities vary from around 2,000 m s^{-1} for some sandstones to 8,000 m s^{-1} for the densest igneous and metamorphic rocks (Sharma 1997: 122). Factors such as fracturing, rock porosity and water content affect wave velocities, thus all these factors can be surveyed by seismological methods. As the speed of a pressure wave in air is only 330 m s^{-1} and in water 1,500 m s^{-1}, water content will generally increase velocity while lower velocities are found in the unsaturated zone. S-waves are not affected by water content as they cannot propagate in water, but for P-waves saturated rocks will exhibit distinct increases in velocity compared to unsaturated rocks. For this reason P-wave velocities are highly useful in hydrological studies. In sedimentary rocks the properties may be anisotropic, with faster velocities occurring in the direction of the bedding planes than across them.

Pressure

Pressure in a fluid is due to the weight of fluid above pressing down. However at any point in the fluid, pressure acts equally in all directions. Pressure is defined as force per unit area. The SI unit of pressure is the Pascal (Pa), with 1 Pa = 1 N m^{-2}.

Average pressure, $p = f/A$

The pressure on an area A at depth h is due to the weight of a column of fluid above it, volume Ah (Figure 4.8). For a liquid of density ρ, the mass of this liquid is $Ah\rho$. Pressure is:

$$p = \text{force/area}$$

$$= mg/A$$

$$= Ah\rho g/A$$

$$p = h\rho g \qquad (4.10)$$

Atmospheric pressure is due to the weight of air above us in the atmosphere. As air has a mass, there is a force due to gravity acting on the air above the Earth's surface. Average atmospheric pressure is measured as 1.013×10^5 Pa at sea level, although its actual value varies according to meteorological conditions from about 0.9×10^5 Pa to 1.1×10^5 Pa. Atmospheric pressure decreases continuously with increasing height. As it does so, the air gets thinner, in other words its density decreases – so there is less weight pressing down for every metre of height. Thus the decrease is not linear, as it is for an incompressible liquid like water, and the formula above cannot be applied.

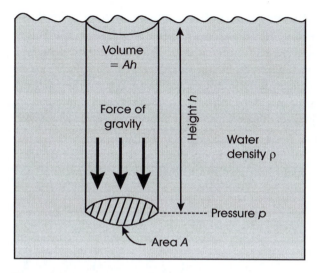

Figure 4.8 Pressure in a fluid.

Atmospheric pressure is often also measured in atmospheres: 1 atm. $= 1.013 \times 10^5$ Pa; or bars: 1 bar $= 10^5$ Pa, so 1 bar or 1,000 millibar is approximately one atmosphere; or mmHg (mm of mercury). Pressure in a mercury barometer is given by $p = h\rho g$, with ρ the density of mercury, which is 13,600 kg m^{-3}.

$$h = p/\rho g$$
$$= 1.013 \times 10^5/(13{,}600 \times 9.8)$$
$$= 0.76 \text{ m}$$

So 1 atmosphere $= 760$ mmHg.

Because pressure depends on the height of overlying fluid, the pressure is the same at the same depth anywhere in a liquid. This explains the principle of a siphon, and movement of water in supply networks and groundwater systems.

For water, with density of 1,000 kg m^{-3}, pressure at depth h of 10 m is:

$$p = h\rho g$$
$$= 10 \times 1{,}000 \times 10$$
$$= 10^5 \text{ Pa}$$

This is approximately equal to atmospheric pressure – in other words every 10 m of depth of water is equivalent to increasing pressure by one atmosphere. This is why snorkels only work on the surface. If you made yourself a snorkel 10 m long and tried to breathe through it underwater, you would rapidly drown, because you would be trying to breathe air at 1 atmosphere pressure with 2 atmospheres pressing on your body, and you would not be able to breathe in.

Archimedes' principle

Archimedes proposed that when a body is submerged in a fluid, there is an upward force or upthrust acting on it, equal to the weight of fluid displaced. So the weight of an object in water is its weight in air less the weight of water displaced. This upthrust force explains why objects feel lighter underwater and why things float.

The origin of the upthrust can be understood by considering the pressure around an underwater object. Pressure at the lower surface will be greater than that at the top, because it is deeper underwater, resulting in a net upward force, countering gravity pulling downwards. The same is of course true for any liquid or gas, not just water. An object's weight underwater will depend on its density. A material with density close to that of water will seem virtually weightless. For steel with density 8.5 times that of water, the weight underwater will only be reduced to 7.5 times that of water. In retrieving objects from the sea floor, the weight of steel cables and steel hulls of shipwrecks will only be reduced by this amount, and thus problems ensue of very long cables breaking under their own weight. An alternative method is to use underwater balloons, attached to the object to be retrieved then inflated, to lift it to close to the surface where a fixed link can be attached.

Flotation

An object floats if the upthrust from the liquid is greater than its weight; i.e. if its density is less than the fluid it is floating in. For a floating object, the upthrust from the submerged portion is equal and opposite to the weight of the object.

An iceberg has relative density of 0.9, and is floating with V_a of its volume above water, V_w of its volume below, and total volume V_i. The force due to gravity on the whole iceberg is balanced by the upthrust due to the volume of water displaced, which is V_w. The volume of the ice below water can be calculated, given density of ice ρ_i and of water ρ_w:

$$\text{Volume:} \qquad V_i = V_w + V_a$$

$$\text{Upthrust force:} \quad F_u = mg$$
$$= -\rho_w V_w g$$

$$\text{Weight of ice:} \quad F_i = mg$$
$$= \rho_i V_i g$$

$$\text{Flotation:} \qquad F_i = -F_u$$
$$\rho_w V_w g = \rho_i V g$$
$$V_w = V_i(\rho_i/\rho_w)$$

So the volume below water is proportional to the ratio of the density of ice to the density of water, i.e. 9/10 of the iceberg is below water – unluckily for the *Titanic*!

It is interesting to note what will happen when an iceberg melts. The meltwater from the ice will have a density the same as water, and so it will contract to a new volume. As the mass remains the same, the new volume (V_{melt}) will be given by:

$$V_{melt} \rho_w = V_i \rho_i$$
$$V_{melt} = V_i(\rho_i / \rho_w)$$

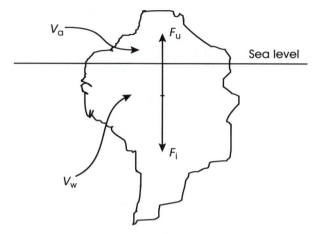

Figure 4.9 *Icebergs.*

This is the same as the volume of ice below water before it melted. In other words, the top bit vanishes but the overall effect on water level is zero. So any melting of the floating ice in the Arctic due to global warming would have no net effect on sea level.

Box 4.4

Scuba diving and smart buoys

Scuba divers adjust their buoyancy by letting compressed air in and out of a buoyancy jacket, to balance their weight plus the weights on their belt with the upthrust force from the water. The aim is to achieve neutral buoyancy, so that the two forces are equal and the diver can float still at a given depth or swim to ascend or descend. When a diver first reaches their required depth, the air in the jacket is adjusted to this neutral buoyancy. However if they then descend further, the increased pressure of the water will reduce the volume of the jacket and thus reduce their buoyancy. The diver has to add more air to retain neutral buoyancy. A fish has a swim bladder that works in the same way. Air is held in it to retain buoyancy – to dive, the fish contracts the swim bladder muscles, and to ascend again it expands it.

In environmental monitoring of deep oceans, underwater buoys that regulate their own buoyancy can be used. These can be adjusted to neutral buoyancy at the required depth. After a period of monitoring (for instance of temperatures, dissolved gases or opacity of the water) they can be programmed to release a cylinder of compressed gas and come to the surface, where the data are transmitted by radio to a receiver. The buoy then lets out the gas and sinks again, repeating the process many times. This principle is particularly useful to measure currents at a set depth underwater as the buoy drifts wherever the current takes it.

Gases

Gases are compressible and have low density because of the lack of interaction between molecules. This means that pressure, temperature and volume are more closely related than for solids or liquids, which in turn means their energetics are quite distinct. As the atmosphere is such a vital part of life on Earth, an understanding of gases is important in climatology and in several aspects of air pollution.

The gas laws

It can be shown experimentally for a gas under pressure, at constant temperature, the volume V is halved when the pressure P is doubled. In other words V is inversely proportional to P, and:

$$PV = \text{constant} \qquad (4.11)$$

This is Boyle's Law.

Further, if the pressure is kept constant and temperature T increases, the volume will increase – like a hot air balloon where turning up the heat expands the gas, increasing the upthrust (from Archimedes' principle). It is found that volume is proportional to absolute temperature, measured in Kelvin (K).

$$V/T = \text{constant} \qquad (4.12)$$

This is Charles' Law. So at room temperature of 27°C, the absolute temperature is 300 K, and increasing the temperature of a balloon by 1°C will increase its volume by 1/300.

The final gas law relates pressure and temperature. For a constant volume of a gas (e.g. an empty bottle), if temperature increases, pressure will increase – so on a hot day, empty petrol cans expand and go clang. If you put a sealed, empty container in the fridge, when you remove it and open it, air will be sucked in to balance the lower pressure inside. In other words, pressure is proportional to absolute temperature, or:

$$P/T = \text{constant} \qquad (4.13)$$

These three laws can be combined to give the **equation of state**:

$$\frac{PV}{T} = r \quad \text{where } r \text{ is a constant}$$

or:

$$PV = rT \qquad (4.14)$$

This is the only equation you really need as it combines all three. The constant r is given per unit mass of the gas as $r = 4.16$ J kg^{-1} K^{-1}, with volume in m^3, pressure in Pa and temperature in Kelvin.

The 'gas laws', as Equations (4.11) to (4.14) are known, are important whenever measurements are made of gases. The equations are true for an 'ideal gas', meaning one where there is very little interaction between molecules. They are a good approximation for most gases such as dry air, CO_2 etc., but they do not hold for some gases with unusual properties, or those close to their freezing points – like water vapour and some hydrocarbons.

Gases are often measured in terms of their molar volume, as one mole of any ideal gas occupies the same volume. A mole is the molecular weight of a substance expressed in grams, which is actually equivalent to a certain number of molecules of any substance – one mole of anything is 6.02×10^{23} molecules. This is Avogadro's Number, N_A.

The equation of state can be rewritten in molar terms as:

$$PV = nRT \qquad (4.15)$$

where n is the number of moles and R is the molar gas constant, 8.31 J mol^{-1} K^{-1}.

Conversion to standard temperature and pressure

To correctly specify a quantity of gas, it is meaningless just to give a volume, as a small change in temperature or pressure would change this volume. You either need to know its mass (which is difficult to measure), or all three of volume, pressure and temperature. To get round this problem, quantities of gases are generally stated as the volume they would occupy at Standard Temperature and Pressure (STP), defined as 273 K (0°C) and 1 atm (101,325 Pa or 760 mmHg).

For instance, an industrial plant is found to emit 2,000 m^3 per hour of flue gas containing 0.1 per cent of sulphur dioxide, at a temperature of 80°C, at atmospheric pressure. In order to find the volume of sulphur dioxide emitted, it must be converted to STP. First convert the temperature to Kelvin, 80°C = 353 K. From the equation of state, Equation (4.14):

$$\frac{P_1 V_1}{T_1} = \frac{P_2 V_2}{T_2} = r$$

where the subscripts 1 and 2 refer to the values before and after conversion to STP.

Rearranging this and putting in the values gives:

$$V_2 = \frac{P_1 V_1 T_2}{T_1 P_2}$$

$$= \frac{V_1 T_2}{T_1}$$

$$= \frac{2000 \times 273}{353} = 1{,}547 \text{ m}^3 \text{ h}^{-1}$$

Thus the volume of sulphur dioxide emitted is 0.1 per cent of this per hour, or $1.547 \text{ m}^3 \text{ h}^{-1}$.

At STP, one mole of any ideal gas occupies 22.4 litres. So the mass of gas emitted could be calculated from its volume.

Work done in expanding gases

When a gas expands, it does work, using energy supplied by the change in temperature or an external application of pressure or heat. In a car engine, energy from petrol heats the cylinder and expands the gas inside: energy is converted from heat to movement in expanding the gas.

This energy is represented by the product of pressure and the change in volume, deducible from the units it is expressed in. Pressure is defined as force per unit area, measured in N m^{-2} (Newtons per square metre), thus pressure $\times$ volume is force per unit area $\times$ volume, with the unit $\text{N m}^{-2} \times \text{m}^3 = \text{N m}$. But in Chapter 2 we saw that one Joule by definition is equal to 1 N m (from the fundamental energy equation, energy = force $\times$ distance) so pressure times volume must also be a measure of energy.

So in any expansion of gas, energy is being transferred between heat and pressure/volume. Compressed gases can therefore be used to store energy, and could represent a useful transport 'fuel'. A vehicle could be powered from cylinders of compressed air, which would be gradually released to drive the motor directly without the need for combustion. The energy stored in compressed gas or air will be given by the pressure times the change in volume:

$$E = P \, \Delta V$$

This is relevant in the Carnot cycle, discussed on pp. 110–11. It can be seen that the pV diagram (Figure 3.6) represents this energy.

Heat capacities of gases

When a gas is heated up, the temperature rise not only depends on the heat capacity and the amount of heat put in (as for a solid or liquid – see Chapter 3), but also on whether the pressure increases, or the gas expands. Because of the gas laws, it will tend to do both depending on circumstances. For this reason it is not possible to define a simple heat capacity for a gas, but instead it has two.

If 1 kg of a gas is heated in a closed container, keeping its volume constant, the heat energy needed to raise its temperature by 1°C is c_v, its specific heat capacity at constant volume measured in J kg^{-1} K^{-1}. Similarly, if the gas is heated at constant pressure, such as in the atmosphere or in a balloon, it has a specific heat capacity at constant pressure, c_p.

Because gases are more readily measured in moles than in kg, these are often expressed per mole of the gas rather than per kg, and are termed the molar heat capacities at constant volume and pressure, symbols C_v and C_p, measured in J mol^{-1} K^{-1}.

If a gas is in a closed container, all the heat applied goes into increasing its temperature. But if it is maintained at the same pressure, when it is heated it will expand, and do some work in this expansion, pushing against the pressure around it. The energy for this work has to come from somewhere, so the heat capacity at constant pressure is higher than at constant volume, i.e. it takes more energy to heat it up by one degree at constant pressure, because it is also using energy in expanding. It is found that the difference between the two is equal to the gas constant R – this can be shown mathematically from the gas laws.

Adiabatic and isothermal expansion

When a gas expands at constant temperature, we can say it is **isothermal**. Pressure and volume change according to PV = constant. However some work is being done in expanding the gas (pushing against a piston for instance), so in order to keep temperature constant some heat or other energy must be supplied. If you squeeze a balloon, and thus reduce volume by increasing pressure, you are doing work and putting energy into the balloon.

If no external energy is being put into the system, changes are known as **adiabatic**. An example of an adiabatic change is when gas is released from a pressurised cylinder, e.g. a diver's tank or a camping gas cylinder. When the cylinder is opened, pressure (at the exit valve) is reduced, and work is being done in expanding the gas as it escapes. This work comes from the heat energy in the gas, and so it will feel cooler.

In an adiabatic change, Equation (4.15), $PV = nRT$, will apply, but the change in pressure will lead to changes in both volume and temperature. In fact, the changes will be governed by the equation:

$$PV^\gamma = \text{constant} \tag{4.16}$$

where γ is a constant equal to the ratio of the molar heat capacities, C_p/C_v of the gas.

In many environmental contexts, changes are adiabatic because no work is being done, and no external energy is being put into the system. If a body of warm, less dense air rises in the atmosphere, the pressure will reduce as it rises. So the body of air will expand according to the lower pressure. In the atmosphere, there is generally no source of external energy for this expansion. The energy must come from the body of air itself, so it will cool. This leads to an equilibrium, as the air will reach a level where it is the same temperature as the surrounding atmosphere.

This is the main reason why air temperature decreases with height, and is known as the **adiabatic lapse rate**. Once in equilibrium air will no longer be rising or changing pressure and volume but be at rest in its natural state, with temperature declining at a set rate with height. For dry air the value of the adiabatic lapse rate is about 9.8°C per km of height.

For moist air, the amount of water vapour the air can hold depends on its temperature. Hot air can hold more water in the form of water vapour, while in cold air the water condenses out. So if moist, warm air rises and cools adiabatically this will cause some water to condense as clouds. When water vapour condenses, it releases energy: the latent heat of evaporation that was needed to form the water vapour. This will warm the air. So moist air rising adiabatically will not cool as much as dry air, but instead some of the energy need for its expansion will come from the clouds that condense out of it. For saturated air, the rate at which it cools (the **saturated adiabatic lapse rate**) is about 6.5°C per km, although it depends upon the air temperature as this determines how much water can be held. Thus the actual rate of cooling will vary between these two rates, depending on the amount of water in the air and the temperature.

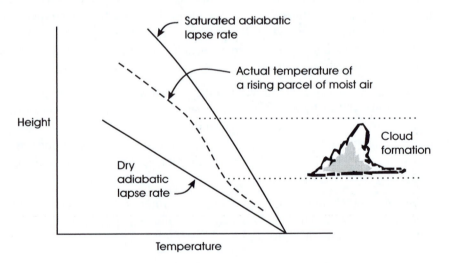

Figure 4.10 *Wet and dry adiabatic lapse rates.*

Fluids and fluid flow

Fluid dynamics has a host of applications, including groundwater flow, sedimentation in streams and estuaries, water supply and treatment systems, the movements of air pollutants in the atmosphere and water, the behaviour of deep sea currents, aerodynamic design of wind turbines, reduction of drag in cars and other vehicles, and flight of birds and aircraft. Flow in a fluid, unless under very simplified conditions (such as in a uniform pipe), can be very complicated to predict or even to measure, so only the fundamental concepts will be touched on here.

Box 4.5

Non-Newtonian fluids

Many naturally occurring fluids are non-Newtonian, including thixotropic mixtures, plastic fluids such as semi-molten rock in the Earth's mantle, or sewage sludge. For these non-Newtonian fluids, viscosity varies with the velocity or shear stress applied, and a non-zero yield stress may apply, the minimum stress needed for the fluid to move at all. This makes mathematical treatment even more complex.

Mix a couple of teaspoons of cornflour (or gravy powder will do) with a few drops of water so the flour is just wet. The resulting mixture is almost solid when you are mixing it – you can mark sharp features with a knife, press down hard on it without it yielding, or pick it up as a lump and throw it from one hand to the other without getting sticky. But leave it still for a couple of seconds and it will quickly all turn liquid, running out into a pool. The forces from being mixed or thrown around act on the mixture to make it cohesive, strengthening friction between flour particles and increasing viscosity. Reduce these forces, and the water flows between the flour particles, reducing friction and viscosity. The mixture is **thixotropic**, having physical properties that are stress dependent. The opposite is observed in tomato ketchup, where shaking the bottle makes the sauce runnier, so it all comes out at once.

Many solid/liquid mixtures exhibit sudden changes in their behaviour from solid to liquid, which are in fact transitions in viscosity due to external force. The slime produced by a snail provides a solid layer, to protect the soft body from sharp surfaces and give traction to move, while turning liquid under pressure to allow the creature to glide along its trail. In this case external force breaks bonds within the slime, allowing it to behave like a liquid. In soils, saturated soil on a hillside can appear quite solid at one moment, but as a result of a slight increase in water content, a vibration or a change in the surface load it can suddenly turn liquid, collapsing as a mudslide with possibly disastrous consequences (as described in Box 1.1).

Similar properties can be made use of in technologies such as fluidised bed combustion (FBC) in electricity generating stations. The fluidised bed consists of a mixture of high-pressure air and pulverised fuel (coal ground to a powder). The mixture will flow like a liquid, allowing combustion to take place efficiently and rapidly under controlled conditions, minimising emissions.

Viscosity is a measure of how treacly a liquid is and thus resistant to flow – a viscous liquid has stronger attraction between molecules, and so when it moves, friction between adjacent fluid elements means that energy is dissipated. One of the assumptions used in most fluid dynamics is that frictional forces increase linearly with the rate of change of velocity of the fluid, with viscosity being the constant of proportionality. This assumption gives the definition of viscosity, and was made by Newton. A fluid that follows this behaviour is known as a Newtonian fluid, such as water.

Compressibility is how much a fluid can be squashed. Most common liquids (water, oil, brake fluid) are virtually incompressible – although they will all compress if squeezed hard enough. Gases are compressible, their volume calculable by the gas laws.

Another quantity used to characterise flow is the **Reynolds number**, which represents the ratio of viscous to inertial forces. The Reynolds number is defined as $\rho v l/\mu$ where a fluid of density ρ and viscosity μ flows at velocity v relative to a solid of length l, and is dimensionless, i.e. it has no units. It is useful in physical modelling and comparing flows, as at the same Reynolds number, two fluids behave in the same way. For instance, the flow of oil in a well could be modelled using water, but adjusting the size of the model to compensate for its different viscosity, keeping the Reynolds number the same. Low Reynolds number (less than 1,000) indicates slow, viscous flow where drag forces predominate over momentum and aerodynamics. For instance while the rapid flight of a bird will have a high Reynolds number (typically around 100,000), for some tiny insects' flight the Reynolds number will be very low due to their size, such that the viscous forces predominate as if the insect were swimming through treacle rather than flying (Jones 1997: 124).

Laminar and turbulent flow

Fluid flow is said to be **laminar** when it is uniform and orderly, but at higher pressures and velocities it becomes **turbulent**. For instance, water flowing along a deep, slow moving river is laminar, but when it goes over a waterfall it is turbulent – flow is no longer ordered, rather the motion is irregular and has many different speeds and directions, containing eddies of various sizes. Turbulent flows contain a degree of chaos that makes calculations of flow rate far more complex and unpredictable.

The Reynolds number indicates whether flow will be turbulent. For flow over a flat surface, turbulence occurs at Reynolds numbers over 5×10^{-5}, while for flow in a cylindrical pipe, laminar flow can be maintained up to Reynolds numbers of 2,200 (Guyot 1998: 96).

In the atmosphere, turbulence is caused in the lower layers at normal wind speeds, with Reynolds numbers of 10^9–10^{12}. Turbulence causes mixing of neighbouring layers, and energy and momentum can be dissipated and randomised, while pollutants are rapidly dispersed. When a plume of smoke emerges from a chimney on a windy day, turbulence can be observed in the atmosphere. The smoke starts off as a coherent column but as it moves away from the chimney, it spreads out and breaks up into discrete chunks, or may disperse in eddies. Very close to ground level however (within a few mm), velocity is much lower due to the surface roughness of the ground, creating a layer with low Reynolds number, around 30. This indicates viscous flow, with much less mixing and processes such as diffusion become important (pp. 143–5).

Measuring flow rates

The **flow rate** of a stream or of water in a pipe is the amount that is passing one point at any time, measured in $m^3 s^{-1}$ or litres per hour, etc. The **flow velocity** is the speed at which it is flowing, in $m s^{-1}$. It is much more straightforward to measure the velocity of flow than the flow rate of a river, but in pollution monitoring you may often need to know the flow rate. For instance if the maximum allowable concentration of a certain pollutant in a river is to be enforced, a polluter may wish to know how much their outflow is diluted by the river water, and thus what concentration can safely be allowed in the outflow. Or an enforcement agency may know what concentration of a pollutant is found in a river in $mg l^{-1}$, and wish to calculate the total load in $kg y^{-1}$ being discharged. In either case, they need to know the flow rate.

Flow rate of a very small stream could be measured by diverting it into a bucket and timing how long it takes to fill it – but this is hardly practical with the Mississippi for instance. Velocity however can be measured with a simple meter that you dip into the river. So it is useful to be able to convert from velocity to flow rate.

In a stream or in a pipe with varying size, with an incompressible fluid flowing along it, the same amount of fluid must pass through any point at the same time, assuming none enters or leaves. In a stream or river water flows slowly where it is wide and deep, and more rapidly where it is shallow or narrow – but the same amount of water passes each point in one second anywhere down its length; likewise in a pipe of varying diameter. This is the **principle of continuity of flow**.

The volume passing any point in one second must be equal to the area at that point multiplied by the distance travelled downstream in one second, in other words the velocity. This volume is the flow rate, and so is given by:

$$Q = vA \tag{4.17}$$

where v = velocity, Q = flow rate in $m^3 s^{-1}$; and A is the cross-sectional area of the flow.

The environmental scientist may thus assess flow rate by measuring the depth and breadth of the river and calculating its cross-sectional area, then finding the average velocity of flow (which may necessitate taking an average of several measurements if flow is not uniform). Flow rate is then found from velocity multiplied by area.

Laminar flow in a pipe

Laminar flow in a pipe, with a pressure difference forcing water through, is one of the simplest flow types that can be imagined, being unidirectional, laminar, and closely controlled, and it is one of the few that can be described by a simple equation. A (microscopically thin) layer of water touching the pipe will not move, and the water closer to the centre of the pipe moves most rapidly. This is due to the frictional forces between water and pipe, and the viscous forces between layers of moving water as the centre of the pipe is approached.

The volume of water passing through per second, or **flow rate**, will thus depend on the radius of the pipe (r), the viscosity (η), the pressure difference (p) and the length of the pipe (l). In fact it can be shown that:

$$\text{flow rate} = \frac{\pi p r^4}{8\eta l} \tag{4.18}$$

This formula is not valid if flow becomes turbulent, at high velocities or pressures.

If flow remains laminar, the equation can be used to calculate the flow rate, or the pipe size and pressure needed to achieve a certain required rate. Because of the fourth power in the radius, a small increase in pipe size would greatly increase flow rate – doubling the pipe size would increase the flow rate by a factor of sixteen, while doubling pressure difference would only double flow rate.

Bernoulli's principle

When a fluid flows, its pressure varies as a result of that motion. It is found that pressure decreases in a fluid where it is moving, and also varies as a result of gravity due to any changes in height. The relationship between the three – pressure, velocity and height – is given by Bernoulli's principle, which states:

> The sum of the pressure at any part plus the kinetic energy per unit volume, plus the potential energy per unit volume there is always constant.

Mathematically this can be stated as:

$$p + \tfrac{1}{2}\rho v^2 + \rho g h = \text{constant} \tag{4.19}$$

where p = pressure, ρ = density, v = velocity, g = gravity and h = height.

This principle can be derived from the conservation of energy. The three terms in the equation denote the three energy types: energy due to pressure, kinetic energy and potential energy due to height. A reduction in energy due to a change in one of these terms must be compensated by an increase in energy due to another. Strictly it applies only to laminar flow of an incompressible, non-viscous fluid (such as water), although it can be applied qualitatively to most fluid motion. In a viscous fluid such as oil, flow causes friction through viscous forces, and likewise with turbulence in turbulent flow, leading to energy loss as heat.

Where movement is horizontal, Bernoulli's principle means that where water is moving fastest, pressure is lowest and vice versa. An example is the suction felt when a fast moving lorry passes a pedestrian or cyclist. Air moving with the lorry travels more rapidly as it passes between the pedestrian and the lorry, resulting in a reduction in pressure. The pedestrian may then be pushed towards the lorry by the greater force of atmospheric pressure acting on their other side.

There are many environmental applications of this principle. It is particularly important in groundwater flow, as this combines the three elements of changes in height, velocity and pressure due to the various rock or soil formations the water flows through.

The **venturi effect** is a specific application of Bernoulli's principle. If a fluid flowing in a pipe passes through a constriction, it will increase in velocity at that point due to continuity of flow. Assuming the pipe is horizontal, this leads to a decrease in pressure at the constriction, due to Bernoulli's principle (see Figure 4.11). This principle is used in a **venturi meter**, an instrument used to measure flow rates. A venturi meter consists of a horizontal pipe with a constriction, connected to pressure meters such that the pressures can be compared to find the flow rate.

The venturi effect is also used in air pollution control, in a **venturi scrubber**. Here flue gases pass through a constriction where they are sprayed with high pressure water. The constriction causes a pressure drop, such that the water and gas are mixed violently in tiny drops and gas bubbles ensuring the maximum contact

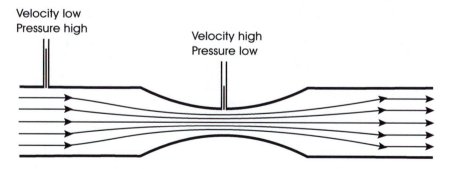

Figure 4.11 *Bernoulli's principle.*

between the two. This allows particulates with sizes greater than around 5 μm to be removed.

Aerodynamics and aerofoils

An aerofoil works by passing a current of air either side of an asymmetric object, such as the cross-section of a wing in Figure 4.12. Because the air has further to travel around the top side, due to its curvature, it must travel faster. By Bernoulli's principle this results in a decrease in pressure. The excess atmospheric pressure on the lower side thus pushes up on the wing, providing the **aerodynamic lift** force.

In flight, the forward motion of an aeroplane combined with the wing's aerofoil shape results in sufficient lift to keep it in the air. This lift is only effective when the plane is moving forwards fairly rapidly however. To take off, it must accelerate rapidly to a high enough speed to gain this lift, which takes a lot of energy and a long runway. Once airborne and moving, flight takes less energy, as there is no friction apart from air resistance, and constant height can be maintained to prevent doing work against gravity. A smooth aerodynamic shape is vital, as at high speed any irregularities would produce eddies and turbulence, increasing air resistance and wasting energy. Turbulence destroys the aerofoil effect, and its prevention is equally important to birds, aircraft and wind turbines.

Jet aircraft are designed to travel at high speed, and have relatively small wings, sufficient to provide lift at this high speed. They could not glide for any distance, and rely upon their engines for forward motion. At the opposite extreme, gliders need much larger wings to give them lift at low speeds, with forward motion provided by gradual descent. Energy is provided by their original towing to get airborne and supplemented by rising on thermal currents. Neither can stand still in the air.

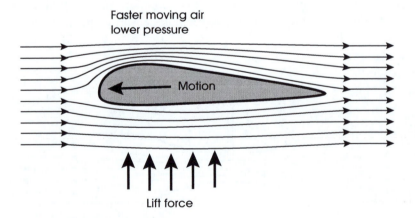

Figure 4.12 *Aerodynamic lift.*

Box 4.6

Bird flight

Birds are supremely evolved to minimise energy needed in flight. Weight is minimised by birds having hollow bones and very lightweight frames, while feathers are a very lightweight insulator compared to fur or fat. Even body organs are kept to minimum sizes to reduce weight. Before a long migration flight, birds stock up on fuel, increasing their body fat reserves to burn during the long journey. Many birds like geese use techniques such as flying in formation to reduce their energy needs in flight. As for aeroplanes, much of the energy is used in take-off. Birds use very strenuous flapping, providing a direct upwards force against the air to get off the ground, sometimes combined with running or jumping.

Like planes, birds have different flying styles. Feathers are each designed asymmetrically as aerofoils, which gave birds a great evolutionary advantage over other flying creatures such as insects, gliding squirrels and bats. Once in the air, they can travel fast and over long distances, using flapping to provide forward motion and the aerofoil for lift. Some birds are great gliders, others have smaller wings and more rapid wingbeats for fast powered flight. Many use the wind or thermal currents to supplement their energy. Like aeroplanes birds are designed to be smooth and aerodynamic to prevent turbulence.

But to hover requires a different mechanism – lift relies on forwards motion. Seabirds appear to hover by flying into the wind, but still gain lift as they move relative to the air while motionless relative to the ground. Kestrels hover by very furious flapping, almost like a Harrier Jump-jet (presumably the name Kestrel Jump-jet didn't sound quite right), but again rely on facing into a breeze to remain stationary relative to the ground. True hovering birds are the humming birds. They do not use aerodynamic lift at all, and in fact their wings are small and not aerofoil shaped, performing figure-of-eight movements as if swimming in the air. This requires large amounts of energy and extremely rapid wing movements to remain airborne, which results in the humming sound – more like a bee than a bird.

Wind turbines also use the same aerodynamic lift force to turn the blades. The blade has an aerofoil shape, producing a 'lift' force that will be at an angle to the blade. This force results in a forward rotation of the blades. A modern wind turbine blade is a very highly engineered precision structure, similar to an aeroplane wing. The tip of the blade moves surprisingly quickly – up to hundreds of kilometres per hour in large machines. One limiting factor in turbine design is that the tip speed must be less than the speed of sound to prevent sonic boom – a limit that otherwise could easily be broken in a large machine at high speed. On a wind farm, the neighbouring turbines must be spaced at large distances such that the wind pattern is not disturbed by an upwind turbine – a distance of ten times their height is considered suitable to prevent one turbine being affected by turbulence from another.

Atmospheric transport of pollutants

High stacks for polluting industries and power stations were introduced in order to avoid localised air pollution such as the London smogs of the 1950s. More recently, this 'dilute-and-disperse' approach to air pollution has been replaced by more absolute measures of emission reductions. However modelling of air pollution transport is still of great importance, and is now addressed to international issues. To reduce forest decline from acid precipitation in Sweden, for instance, will it be more effective to reduce emissions in Scandinavia, in the UK or in eastern Europe?

The movement of pollutants in the atmosphere takes place over timescales that can vary from seconds to years, and over distances from a few tens of metres to global air pollution extending thousands of kilometres. While horizontal movements are chiefly due to wind, vertical movements are generally slower, as a result of convection and turbulence. For a pollutant to reach the stratosphere may take years, and once there it may not be removed for centuries. At lower levels there is more turbulent mixing, water and water vapour are present, and rain to dissolve and remove the pollutant. As the fluid dynamics of plumes is highly complex, with chemical processes also important, no mathematical treatment is attempted here. More advanced texts such as Boeker and van Grondelle (1999) or Seinfeld and Pyros (1998) are recommended.

The simplest mathematical model of air pollutant transport is of a Gaussian plume. Essentially, this assumes that the pollutant is drifting downwind at the prevailing wind speed, whilst simultaneously spreading out in all directions at a rate depending on turbulent mixing and the concentration of the pollutant. A model based on a statistical distribution can be developed, similar to Fick's Law in Equation (4.1) for molecular diffusion, but in three dimensions, and with more rapid movements associated with bulk flows. Turbulent diffusion causes the pollutant to disperse down the concentration gradient in the same manner as molecular diffusion, but at a much faster rate, driven by turbulent motion of the atmosphere rather than random molecular motion.

From this model the typical concentration of pollutant at ground level can be calculated around the stack. One interesting outcome is that the maximum ground level concentration is found to be inversely proportional to the square of the stack height, justifying the construction of tall stacks. This will be found at a distance depending on height and dispersion – the more turbulence, the closer it will be to the stack. This type of model can be used in stack design by predicting concentrations, and can explain observed patterns, for instance the concentrations accumulated in soils of pollutants from a specific source (Figure 4.13). The actual behaviour however depends not only on height and wind speed but on other factors determining turbulence such as ground surface roughness and atmospheric effects.

Plate 4.2 *Smoke plume dispersing from an industrial chimney in New Jersey.*
Source: photo by courtesy of Professor P. Brimblecombe, University of East Anglia

Wind speed increases with height, due to friction with the ground surface. Surface roughness affects this increase, so for instance the wind speed increases more rapidly with height above open water than above the canopy of a forest. Turbulence is associated with higher wind speeds, large temperature gradients and higher surface roughness, but decreases with height as the air is further away from the

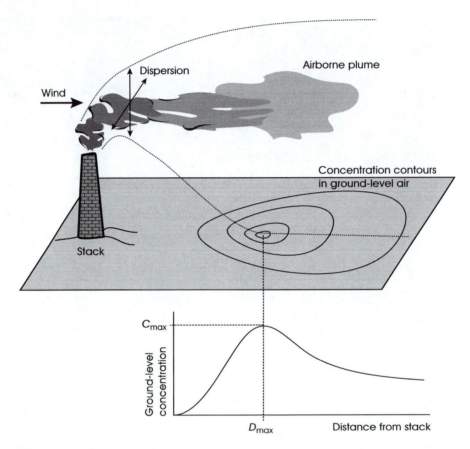

Figure 4.13 *The Gaussian plume.*

main cause of turbulence, i.e. friction with the ground. Above the **boundary layer**, which may be at a height anywhere between 100 m and 3,000 m, winds are no longer affected by the ground surface but purely by atmospheric factors. Winds also depend on surface features such as mountains which can create updraughts, and increase both wind speeds and turbulence around them.

A higher stack will not only mean the pollutant is released further from the ground, but it reaches a level of higher winds, and it is hoped less turbulence. Under stable conditions the pollutant may then travel hundreds or even thousands of kilometres before it is removed from the atmosphere by rain, during which time it will disperse both horizontally and vertically to lower concentrations and may fall relatively harmlessly over the ocean rather than on land.

As smoke is emitted from a stack, its temperature is higher than the surrounding air and it has some vertical momentum, so its initial movement is upwards. It will be moved horizontally in the direction of the wind, and as it cools its course will level out. Past this point, it may continue in a level stream gradually spreading out, it

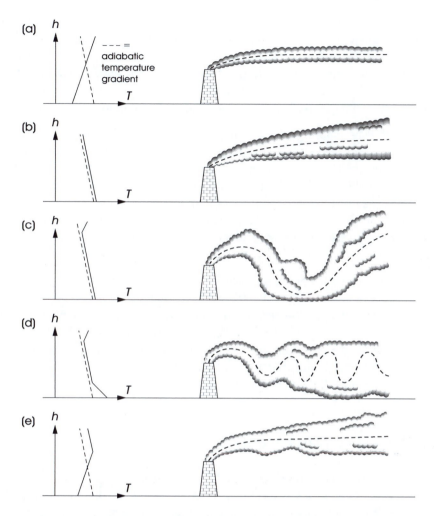

Figure 4.14 *Plumes and stability of the atmosphere: (a) stable from ground up to above chimney; (b) neutral from ground up to above chimney; (c) stable above chimney, neutral at ground level; (d) stable above chimney, unstable at ground level; (e) neutral above chimney, stable below.*
Source: reproduced by permission of KNMI from van Dop (n.d.) p. 18

may rise or drop to ground level, or it may fluctuate turbulently. These different plume shapes depend upon wind and stability in the atmosphere, that is the tendency to damp or amplify any vertical movements of the plume.

Stability is related to the adiabatic lapse rate of the atmosphere (p. 160), and describes the behaviour of a parcel of air with slightly different properties to its surroundings. As a parcel of moist air rises, pressure decreases and it expands, reducing its temperature adiabatically. If moisture condenses out it will not cool as much, as latent heat will be released. In a stable atmosphere, as the parcel of air rises, it cools faster than its environment and thus becomes less buoyant, returning it to its original position. This results in a level flow of the plume (Figure 4.14(a)). Under unstable conditions, the parcel cools less than its environment, making it more buoyant, which causes it to rise faster, while if it starts to fall, it becomes less buoyant and falls further. Unstable conditions lead to the plume dropping to ground level, or travelling in vertical waves of increasing amplitude (Figure 4.14 (c), (d)).

The actual temperature gradient, and hence stability, differs from the adiabatic lapse rate as a result of the sun heating the ground, wind moving air of different temperatures, and variations in moisture content. Under hot sun the air near the ground will be heated, resulting in a higher temperature gradient which creates instability, as the adiabatic cooling of the plume will be less than this temperature gradient. On a still, clear night, the ground cools radiatively resulting in a low or negative temperature gradient and a stable atmosphere, with less vertical mixing.

In the extreme is a temperature inversion, where warm air overlies colder air with no vertical mixing, generally in a closed basin and accompanied by fog or smog. In a valley basin like Los Angeles or London, a layer of cold smoggy air can become 'trapped' underneath warmer air above. The situation can occur naturally, as observed in autumn morning mists in valleys, but the presence of particulate pollution exacerbates the situation greatly, as the Sun's warmth cannot penetrate to evaporate the mist and warm the air at ground level.

Surface tension and surface effects

At the surface of water (or other liquids), molecules form bonds with other water molecules, but not with the air above. Within the body of the liquid, attractive and repulsive forces act equally in all directions, but at the surface there is a net force at and parallel to the surface. This results in the relatively strong surface of water, as breaking the surface means breaking these bonds. This is known as surface tension.

Surface tension is responsible for droplets being round, as the surface of the drop is smallest if it is a sphere. A falling drop is an almost perfect sphere as it is in free-fall with gravity; a drop on a solid surface will be distorted by gravity flattening it and may spread out flat, wetting or soaking into the surface. This depends upon the inter-molecular forces between water and the solid on which it lies. On smooth varnish or an oily surface, water will retain its droplet shape. On clean glass, the surface tension will be broken and the drops spread out, as the forces between water and glass molecules are greater than those within the water droplets. A porous surface is one where water experiences forces with the material, such as paper or concrete, breaking surface tension and allowing water to infiltrate the pores.

Other liquids display the phenomenon, for instance mercury if spilt on a smooth table or floor will form spherical droplets, as mercury is not strongly attracted to molecules of any other substance.

The magnitude of surface tension depends upon the liquid, the substance at which a boundary is being made and the temperature. At 20°C, for water in air, surface tension results in a force of 7.26×10^{-2} N m^{-1} (newtons per metre), while for mercury the figure is 46.5×10^{-2} N m^{-1}, and for olive oil and water it is 2.06×10^{-2} N m^{-1} (Nelkon and Parker 1994).

Increasing the area of a liquid's surface requires work to be done against surface tension, and hence a liquid's surface has potential energy. Thus increasing the surface area of a liquid will generally result in a decrease in temperature to supply this energy, known as surface energy. When a liquid evaporates and has no surface it has no surface energy, so the latent heat of evaporation must supply the necessary surface energy.

Surface tension shapes the **meniscus** at the surface of water. In a glass tube, rather than being flat, it is curved and raised at the edges. Again the shape depends upon the liquid and the material – mercury in a glass tube will be curved downwards at the edges. The angle of contact between liquid and glass is determined by inter-molecular forces. For clean glass and water it is 0° while for mercury it is 137° (Nelkon and Parker 1994), indicating that water and glass are strongly attracted while mercury and glass are not. This meniscus explains capillary action, the process by which surface forces between water and glass can pull a column of water up a narrow glass tube (see Figure 4.15). At the boundary with the glass, where the water's surface is vertical (angle of contact = 0°), the force between glass and water must be equal and opposite to the force of surface tension, by Newton's law of action and reaction (pp. 6–7). The surface tension is given by γ, and the length of this boundary is given by $2\pi r$ where r is the diameter of the tube. This force around the circumference of the tube must be equal to the gravitational force pulling the water downwards. As the volume of the water is $\pi r^2 h$, given density ρ and gravity g this gives:

$$\text{Force} = 2\pi r \gamma = \pi r^2 h \rho g$$

$$2\gamma = rh\rho g$$

$$h = 2\gamma / r\rho g \qquad (4.20)$$

Thus the height of the capillary column is inversely proportional to the radius of the tube, for water in a glass tube where the angle of contact is 0°. For a wider tube, the weight of the water is proportionately greater compared to the length of the water/glass boundary and so the column height is lower. Putting in numerical values gives $h \approx 15/r$, with h and r both in millimetres.

Capillarity works in porous materials such as blotting paper in the same way as in straight glass tubes. Capillary action is important in water movements in soils and

(a)

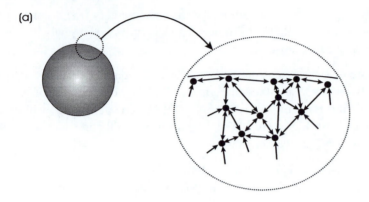

(b)

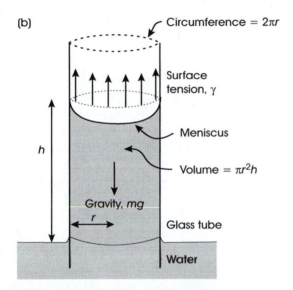

Figure 4.15 *Surface tension and capillarity: (a) in a droplet; (b) capillarity in a tube, showing angle of contact and forces acting.*

rocks, as the pore structure acts as fine capillary tubing and draws water up above the water table.

Water movement in plants also depends upon surface forces although in this case capillary action alone would not be sufficient to raise water to the heights required, to the tops of trees. Water is drawn up long, continuous pipe-like structures made from many dead cells end to end, that are full of water. The column will not break because water has a high tensile strength, with the energy to lift the water coming from evaporation at the leaf pores.

Hydrology and hydrogeology

Hydrology is the study of water in the environment, including seas, rivers, streams and groundwater, while hydrogeology studies geological aspects of water, particularly water underground. Water movement determines the transport of waterborne pollutants, their dispersal or dilution, and the rate at which contamination may spread to cause a potential hazard. Underground flow rates cannot be measured directly, and features such as rock type and characteristics, permeability and cracks that determine water movements are not visible from the surface, so computer modelling is often used to predict flow rates. This demands an understanding of the theoretical factors affecting flows. While hydrology is a vast and complex topic, it is hoped here to give a taste of the key factors to allow further study. Price (1985) and Kiely (1997) are useful texts.

Hydrological processes

If you dig a hole in the beach, you will generally reach water at a level above that of the sea. This level is the water table. It is above sea level because the water is flowing down into the sea. Sand on the beach contains air in the gaps between the grains (or pores), and it is through these pores that water can flow. Similarly, crustal rocks contain pores that may hold water, known as **groundwater**.

Of all the water on Earth, over 97 per cent is sea water, while groundwater comprises 97 per cent of unfrozen water outside of the oceans, leaving only a fraction of a per cent in all rivers, lakes and atmosphere (Price 1985). It is evident that groundwater is of great importance in the provision of fresh water for mankind's use, as well as water movement in the hydrological cycle. Many parts of the world obtain water supplies from boreholes into groundwater. Not all groundwater is readily usable however, as much is either saline or is inaccessible due to its depth.

The amount of water a rock can hold is determined by its **porosity**, which is a measure of the pore volume. Porosity is defined as the ratio of pore volume to rock volume, and may vary from less than 1 per cent for unweathered metamorphic or igneous rocks, anywhere from 5 to 40 per cent for chalks depending upon age and compaction, up to 85 per cent or more for certain laval rocks such as pumice. Highly compacted rocks, and those deep below the Earth's surface may have zero porosity. Weathering can increase porosity in the form of fine cracks, created as water dissolves minerals and increased by the freeze-thaw action of water permeating into them, freezing and expanding. Soil is also porous, with both water and air contained in the pores.

The ease with which water can move through a rock is measured by its **permeability**. Permeability depends upon the structure of the rock including the pore size and how well pores are connected, the shape of grains and the existence of cracks. A rock such as a chalk could have high porosity, but low permeability due to its small pore size. By contrast a weathered granite could have very low porosity, but contain cracks and fissures allowing water to move readily and so have high permeability. Aquifers form in rocks such as sandstone where both porosity and permeability are high. In any rock or soil, water is held in a thin film around rock particles by surface effects, which means that not all the water held in an aquifer can be extracted. The amount held depends upon pore size or soil grain size, and this effect is very important in holding moisture in soil for plants to use. In soils, clays are relatively impermeable because of their small grain size, holding water in their small pores through surface forces, while a sandy soil could have lower porosity but higher permeability.

Figure 4.16 illustrates some hydrological features, in a system where water is trapped below a layer of impermeable rock or clay (the **aquiclude**). This water forms a confined or artesian aquifer. If a well is drilled into the aquifer, water is forced up the well by the pressure of water farther up the hill. Below the water table lies the **saturated zone**, where all rock pores are filled with water, while above it is the **unsaturated zone**. In the unsaturated zone rocks may be damp,

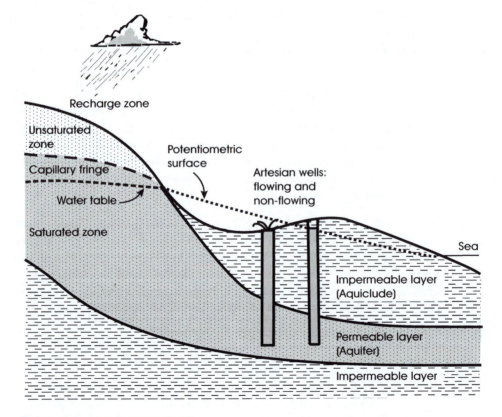

Figure 4.16 *Hydrological features.*

with some water in their pores and around rock grains, but not filling all pores. Between the saturated and unsaturated zones lies the **capillary fringe**, where rocks are saturated due to capillary action, but if a well were dug here it would be dry.

The **potentiometric surface** is the surface level of water in the well or the unconfined aquifer, or for a confined aquifer it is equivalent to where the surface of the water would be if it weren't trapped below the aquiclude. It slopes because water is flowing downhill into the sea, just like the water table under the beach. A flowing artesian well is one dug where this surface is above ground level – water then is forced naturally up the well as the pressure at the bottom is sufficient to do so, due to water farther up the hill in the aquifer. Where this surface is below ground level, water needs to be pumped from the well.

Pressure in an aquifer depends upon depth and velocity of water flow, due to Bernoulli's principle (pp. 164–5). Pressure is generally expressed in terms of effective head, in metres, which can be measured as the height of water in a borehole. Pressure due to the depth of water is given from $p = \rho g h$, where h is the height to the potentiometric surface, plus atmospheric pressure at the surface which is generally ignored as it is constant. In a dynamic situation this pressure will be reduced due to the kinetic energy of the water, although in many groundwater situations this kinetic component is negligible. As water flows through rock however it loses energy through friction, which results in reduced pressure. This is the reason for the slope of the potentiometric surface, and the reduction in pressure is termed **head loss**. The gradient of the potentiometric surface, termed the **hydraulic gradient**, depends upon the permeability of the rock and will determine the flow rate.

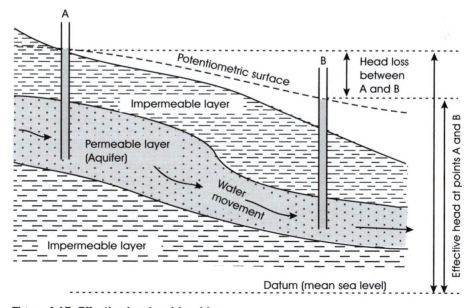

Figure 4.17 *Effective head and head loss.*

Darcy's law

Darcy, a nineteenth-century French water engineer, devised a general law for water travelling through a permeable medium. Darcy's law states that:

$$Q = KAh_L/l \tag{4.21}$$

where Q is the flow rate, K is hydraulic conductivity (or permeability), h_L is head loss, A is cross-sectional area and l is the distance travelled. In other words, flow rate is proportional to the area flowed through, the head loss and the inverse of the distance, with K being the constant of proportionality, units m s^{-1} or m day^{-1}. It is an empirical law, found from observation, and not easily provable even for the simplest case of a homogeneous medium.

This law is analogous to similar equations governing flow of electricity, with the electrical conductivity taking the place of hydraulic conductivity, and the voltage taking the place of head loss. Conduction of heat takes the same form, in this case dependent upon thermal conductivity and temperature difference.

Hydraulic conductivity depends upon both the rock or soil type and the liquid flowing, as opposed to permeability which is dependent only upon the rock type. It is evident that the flow rate for a viscous liquid would be slower than that for water, which would be reflected in a lower value for hydraulic conductivity. This would be relevant in oil extraction for instance, where viscosity varies with oil type,

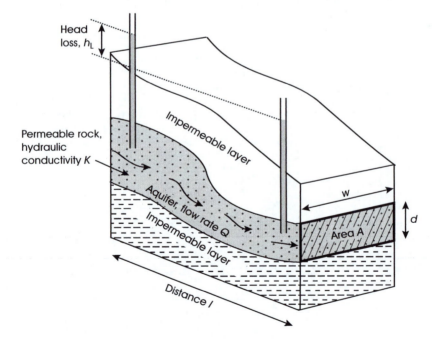

Figure 4.18 *Darcy's law applied to a regular aquifer.*

temperature and so on. However hydrologists who always talk about water, with a constant viscosity, generally find the use of permeability as a function of the rock more useful.

Darcy's law is only applicable to laminar flow, as at higher flow rates when turbulence occurs greater pressure is required to maintain flows, and in fact the flow is found to become proportional to the square root of head loss. However in most natural situations flow rates are low enough for Darcy's law to be applicable.

Groundwater flow

Predicting groundwater flow rates is relevant to water supply, to establish how much water may be extracted from an aquifer on a sustainable basis, and equally so in geothermal power production (pp. 111–13). It is also important in studying movements of contaminants dissolved in groundwater where these might pollute water supply, discussed in the following section, or similarly movements of dissolved nutrients needed for plant growth.

Darcy's law can be applied to predict flow rates through aquifers. The head loss can be measured by taking pressure measurements at different points and, from a knowledge of the permeability of the rock type and the depth and area over which it extends, the flow can be estimated. For a confined aquifer of regular thickness d, over a width w, the head h_1 and h_2 may be measured by boreholes at two points, a distance l apart, to find the head loss h_L (as in Figure 4.17). Given hydraulic conductivity K, the flow through the aquifer can then be found from Darcy's law (Equation (4.21)) as:

$$Q = Kdwh_L/l$$
$$= Twh_L/l \tag{4.22}$$

Here T is the transmissivity, defined as hydraulic conductivity multiplied by depth (Kd) of the aquifer. This has units of hydraulic conductivity multiplied by depth giving $m^2\,day^{-1}$ or $m^3\,day^{-1}\,m^{-1}$, a convenient measure of how much water an aquifer can supply.

For an unconfined aquifer the calculation also takes into account the slope of the water table, which results in the area decreasing and some vertical component to the flow. In real-life situations rocks are not homogeneous, and an aquifer may contain overlying rocks of different permeabilities and depths. Transmissivity of each layer may then be calculated, with the separate values added up to obtain overall transmissivity.

The situation is more complex where fissures carry much of the flow, such as in weathered igneous rocks, as flow is then not homogeneous and far more difficult to quantify. In some rock types such as limestone there may be underground rivers and lakes, that will account for large water movements. Aquifer thickness may vary at different points, and flow may not be in a straight line. In addition many rocks have different properties in different directions, termed anisotropy – for instance a layered sedimentary rock may have higher permeability horizontally than vertically.

The amount of water that may be extracted from an aquifer depends upon the water balance for the region. Aquifers are recharged from rainfall and from lateral movements over a wider area, and in equilibrium this recharge rate will be equal to the rate of water leaving the aquifer from abstraction, natural springs and streams and by evapotranspiration. In some cases an aquifer may contain a large volume of water but have a low recharge rate, where recharge is from a long distance away or permeability is low. Under these circumstances over-abstraction is more likely, as has happened for instance in the artesian basin underlying London where abstraction rates have exceeded recharge for over a century, resulting in a lowering of the water table.

If a well is bored into an aquifer, the maximum pumping rate will depend upon the flow of water through the aquifer into the well, calculable from Darcy's law. As water is abstracted from the well, the pressure here drops, and results in a lowering of the potentiometric surface. This causes flow inwards from the aquifer, at a rate determined by the hydraulic gradient. Around the well, a **cone of depression** forms, or a hollow in the potentiometric surface, to provide the head loss for this movement (Figure 4.19). In the case of an unconfined aquifer, this will result in a lowering of the water table, which will reduce the flow rate as the area of flow will be smaller. If the pumping rate from the well is higher than that sustainable, the well level will decrease until it dries out.

The height of water in a well determines the amount of energy needed for pumping, to supply the potential energy to raise the water to ground level. Well height depends on permeability and flow rate as this affects the shape of the cone of depression. The level in the well when pumping is generally found to be lower than the static potentiometric surface around it, due to losses in energy as the water flows into the well. These come about as the water flows through the sides of the well, which may be lined or compacted, and as the higher speed of the water represents kinetic losses. These effects combined with the cone of depression are known as **drawdown**. Where neighbouring wells penetrate the same aquifer, interaction between their cones of depression can reduce flow further. This has happened in parts of northern India, where the provision of diesel-powered deep tubewells for irrigation has in many instances caused the older shallow tubewells to dry up.

In coastal areas, the groundwater under the sea will be saline. Because salty water is denser than fresh, together with the general tendency for groundwater to flow towards the sea, the saline water is generally kept separate from the fresh. However extraction of groundwater may reverse this trend. A lowering of the water table

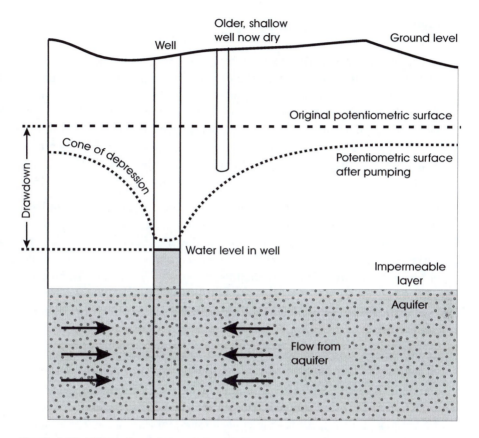

Figure 4.19 *Tubewells and over-abstraction.*

inland and reduced flow towards the sea can result in saltwater intrusion and salination of groundwater. This is the situation in parts of California, where extensive irrigation has depleted the groundwater resource.

Evaporation of water from the surface, and evapotranspiration from plants with their roots drawing water from the soil represent major components of the hydrological cycle. In hot arid climates, evaporation may exceed rainfall over the course of the year, whereas in moist, temperate zones the converse is true. Thus in the case of the arid climate, the net movement of water may be upwards through the soil into the air, and inland lakes may have no outlet, but lose their water through evaporation, increasing their salt and mineral content in the process. This can increase salinity in groundwater, as soluble minerals are moved upwards with evaporation, rather than being leached away downwards. Increasing use of water for irrigation or the building of dams will increase evaporative losses, reducing groundwater levels and river flows correspondingly.

Contaminant transport in groundwater

Where contaminants are dissolved in groundwater, they will be transported along with it as it flows. However there are several other factors determining transport of contaminants. For an insoluble liquid such an oil spillage the situation is more complex. Soils and rocks will also interact with pollutants, which may be adsorbed on to soil particles or held in organic matter, slowing their motion through the soil. Much of a substance may not be in a soluble form and therefore not be mobile or available. The chemistry of these systems is complex, and beyond the scope of this book. Fuller treatment is given in O'Neill (1998).

A point source of a soluble pollutant will not only move with groundwater flow, but also spread out into a wider, more dilute plume, known as dispersion. As groundwater flows through convoluted pathways of pores rather than through direct channels, the contamination collides with and bounces off the pore walls, dispersing it over a wider area. This is termed hydrodynamic dispersion. Diffusion will also play a part, as dissolved substances will diffuse through groundwater and spread out. Mathematically, diffusion and hydrodynamic dispersion can be treated in a similar way.

In the case of a one-off spillage of some hazardous soluble material, on to the soil or into water, both the pattern of dispersal and its rate can be modelled. Over time the pollutant will move with groundwater flow (modelled using Darcy's law, Equation (4.21)), and spread over a wider area with dispersion (according to Fick's law, Equation (4.1)), so that the concentration will decrease. These models together with monitoring data can determine the point and time at which the pollutant level is safe again – a dilute-and-disperse approach.

For more hazardous or long-term pollutants where the dilute-and-disperse approach is insufficient, containment may be more applicable. Low permeability material such as clay or plastics can be used, for instance in lining of landfill sites, to prevent leachate percolating downwards into the groundwater.

In remediation of contaminated land, a combination of approaches is generally used. The most contaminated topsoil may be removed and disposed of in a contained way in landfill, while on-site containment may be facilitated by constructing a clay cap to prevent subsurface pollutants finding their way to the surface. Soil cleaning or bio-remediation may be used where economically viable. The concentrations, movements and availability of remaining pollutants should be monitored, to ensure that they are at sufficiently low concentrations not to present an unacceptable risk. Land uses may be restricted, such that housing or other sensitive uses are kept at a certain distance. Modelling of contaminant transport processes can be a valuable aid in ensuring the safety of the site at least cost.

An increasing problem is the contamination of groundwater and drinking supplies with nitrates from agriculture. As the use of fertiliser is widespread and continuous, any residue unused by the plants makes its way into the groundwater and rivers. The rate at which this occurs depends chiefly upon soil and rock porosity and permeability, which in some cases can delay any problems for decades or centuries. In the long term however the only solution is control at source – a reduction in fertiliser use, or the use of slow-release or organic fertilisers, such that more of that applied is taken up by the plant.

Summary

- Matter may exist as solids, liquids, gases or plasma, according to temperature and pressure, which define their fundamental physical properties. The various mixtures and solutions possible will have properties dependent upon the constituents.

- Condensation nuclei are required for a gas to condense or a liquid to freeze, controlling these processes in the atmosphere.

- The behaviour of rocks and other solids depend upon their strength and elasticity. These features determine wave motion, used in seismology to study sub-surface features.

- The 'gas laws' can be summarised as the Equation of State, $PV = nRT$, which relates pressure, volume and temperature in a gas. Energy exchanges, characterised by adiabatic and isothermal changes, determine the observed values of these variables, important in atmospheric physics.

- Flow rates, turbulence and diffusion can all be important in water and air pollution, acting to dilute and disperse pollutants.

- The main physical principle applied to groundwater flow is Darcy's law, which states that flow rates are proportional to pressure difference and depend on permeability of rocks.

Questions

1 In an irrigation well it is found that the surface of the water is 5 m below the ground, and the well is 30 m deep. What is the pressure at the base of the well? What simple geological situation would explain these features of the well?

2 Why can whales grow bigger than elephants?

3 The Environment Agency measures the average velocity of flow in a river to be 1.1 m s^{-1} where it is 0.5 m deep and 5 m wide.

   (a) What is its flow rate?

   (b) Just downstream, in a narrower part of the river it is 2m wide and 1 m deep. What is the flow rate and the average velocity here?

   (c) The concentration of a toxic pollutant coming from a paint factory at this flow rate is 50 per cent of the EC safe limit. During a drought, the river is observed to slow down to 0.8 m s^{-1} and is shallower at 0.3 m, though still 5 m wide. What is the new flow rate? Is the EC limit now exceeded?

4 A bungee jumper jumps off a bridge 75 m high. As he leaps, he falls with gravity for 40 m until the bungee is taut, and then slows down as the bungee stretches until his hair just dips in the water of the river below, when he starts to bounce up again. After a few bounces he comes to rest with the bungee stretched to 55 m.

The bungee has a cross-sectional area of 0.001 m^2, and the jumper has a mass of 75 kg.

(a) What happens to the energy in the system while this is happening?

(b) What are the stress and the strain in the bungee at final equilibrium?

(c) Assuming extension is linear, what is Young's modulus for the bungee?

5 A scuba diver can withstand pressures up to 4 atmospheres without risk of getting the bends. What is the maximum safe diving depth?

6 A water main of radius 10 cm runs from a reservoir on a hill down to a town, where houses are connected to it by supply pipes of 1cm radius. When one house only switches on a tap, the flow rate is 0.25 l s^{-1}. What is the average velocity of water in the mains, and in the household's supply pipe?

Where is the pressure highest? and lowest?

(a) at the top of the hill;

(b) in the mains outside in the street;

(c) at the tap which is switched on.

Answers to numerical parts

1 351 kPa

3 (a) 2.75 $m^3 s^{-1}$ (b) 1.375 m s^{-1} (c) 1.2 $m^3 s^{-1}$

4 (b) 750 kPa; 0.375 (c) 2,000 kPa

5 30 m

6 7.96×10^{-3} m s^{-1} ; 0.796 m s^{-1}

Further reading

Environmental Physics (2nd edn). E. Boeker and R. van Grondelle. John Wiley & Sons, Chichester, 1999. Includes mathematical treatment of transport of pollutants, amongst a wide range of topics in environmental physics.

Environmental Engineering. G. Kiely. McGraw-Hill, New York, 1997. Includes chapters on transport of pollutants in air and waters, and hydrology.

Environmental Chemistry (3rd edn). P. O'Neill. Blackie, London, 1998. Definitive text on environmental chemistry, in air, water and soils.

Introducing Groundwater. M. Price. Chapman & Hall, London, 1985. Probably everything you will ever need to know about groundwater, in a concise and very readable volume.

Environmental and Engineering Geophysics. P. V. Sharma. Cambridge University Press, Cambridge, 1997. Chapter 4 covers seismic surveys.

Atmospheric Chemistry and Physics: From Air Pollution to Climate Change. J. Seinfeld and S. N. Pyros. John Wiley, New York, 1998. Detailed coverage of air pollution and pollutant transport.

5 The Earth's climate and climate change

The current concern over climate change due to the release of CO_2 into the atmosphere is arguably the single most serious environmental issue we face. The problem is intractable, as large reductions in CO_2 emissions appear politically and economically unacceptable, at any rate in the short term. There is much uncertainty over the precise effect of these emissions, but the warming that we are already seeing is likely to be the start of major climatic changes and sea level rise. In the most optimistic forecast these will require large changes in agricultural practices, investment in sea defences and will result in increased storm damage in the future. The more pessimistic view is that the future ability of the Earth to sustain life will be reduced. The following key concepts are covered in this chapter:

- **The climate system is driven by solar radiation, which creates convection currents that form the general atmospheric circulation**
- **This circulation transfers energy around the globe, and determines conditions that give rise to most weather features we experience**
- **Ocean currents are also important in energy transfer and climate**
- **The ozone layer protects us from harmful UV, but has little impact on climate**
- **Anthropogenic changes to atmospheric composition, principally through emissions of CO_2, are changing climate and increasing mean temperatures – the 'greenhouse effect'**
- **Climate change will include changes in rainfall and the prevalence of storms or other extreme events, and sea level will rise**
- **While there are many uncertainties surrounding climate change, modelling studies have been validated with sufficient confidence for the majority to believe that climate is changing significantly, and that large cuts in emissions will be needed to limit this change**

The Earth's climate

When we refer to the **weather**, we mean the day-to-day variations in temperature, precipitation, pressure, wind and cloud cover. More generally **climate** denotes average weather over a period of years. The Earth's climate is subject to natural cycles and variability over the short and long term, and long-term trends which are due to both natural and anthropogenic factors.

Climate is driven by the energy input from the Sun. This energy heats the Earth's surface and atmosphere, and drives convection currents in the oceans and atmosphere that serve to redistribute the energy around the globe. The atmosphere plays a vital role in absorbing much of the Sun's radiation before it reaches the surface, and in absorbing infra-red (heat) radiation emitted from the Earth. This natural greenhouse effect maintains temperatures within the range we have evolved to tolerate. Water in the atmosphere, in the form of vapour and clouds, also has a key role in affecting both the absorption of radiation and the energy balance. The details of climate are immensely complex and delicately balanced, being dependent upon many feedback effects between atmospheric composition, ocean temperatures and circulation, plant and animal life.

The atmosphere

The atmosphere can be divided into four layers, as illustrated in Figure 5.1. The layer at the surface of the Earth is called the troposphere. It is about 11 km thick (varying with latitude), and it is in this layer that most climatic processes (clouds, winds, rain) happen. The main gases in the troposphere are nitrogen (N_2, 78 per cent), oxygen (O_2, 21 per cent), argon and other inert gases (1 per cent), and carbon dioxide (CO_2, 0.03 per cent). It contains a varying proportion of water vapour

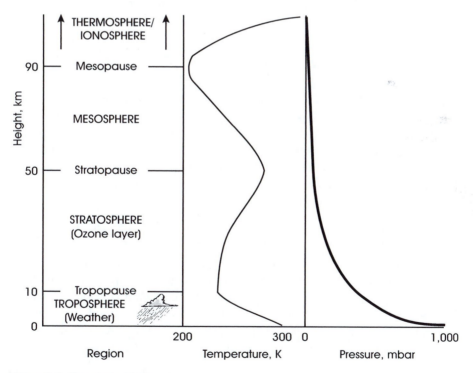

Figure 5.1 *The atmosphere.*
Source: adapted from Critchfield (1983) p. 11

(0.5 to 4 per cent), and may also contain pollutants such as carbon monoxide, sulphur and nitrous oxides, particulates and ozone. The temperature near the Earth's surface is high as the ground heats up from absorbing the Sun's heat and then warms the air near it – temperature then declines with height.

Pressure declines with height throughout the atmosphere, as there is less air above pressing down. This means density also decreases with height. Pressure is roughly inversely proportional to the log of height.

The next layer is the stratosphere, which is about 40 km thick. In the stratosphere there is very little water vapour, and there is some ozone (O_3) formed from oxygen when it encounters high energy radiation from the Sun. This is the 'ozone layer' protecting us from UV. Temperature increases with height, due to absorption of solar radiation in these processes, which produces heat. There may also be some pollutants, including ozone-destroying CFCs, which will be persistent as they will not be washed out by rain, and there is very little mixing between the layers.

Above the stratopause is the mesosphere (40 km thick) and the thermosphere/ionosphere which carries on getting progressively less dense out to 500 km, although there are some sparse particles right out to 80,000 km. These layers together contain only 1 per cent of the mass of the atmosphere mainly in the form of scattered ionised particles, and although they are important in protecting the Earth from high energy radiation, they play little part in climatic processes.

General circulation of the atmosphere

On the Earth, the Sun's energy causes convection currents, in other words winds and ocean currents. These follow regular patterns, determined by the amount of energy available, the depth of oceans and positions of mountain ranges, and also the rotation of the Earth. Winds and currents transfer energy around the globe, and serve to redistribute heat from equatorial to polar latitudes. They also carry moisture, providing rain and the basis of the hydrological cycle, and transferring more energy as latent heat.

The major wind systems, or **general atmospheric circulation**, derive from the hottest air at the equator rising and moving away from the equator at altitude, drawing in colder air at sea level. Rather than circulating all the way to the poles, the movement is broken up into three cells, at the tropics, mid-latitudes and poles (Figure 5.2). As air is drawn to the north and south it is also deflected by the rotation of the Earth by Coriolis forces (p. 21), producing the dominant north-easterly and south-westerly winds in the northern hemisphere. For a more detailed treatment of these and other climatic features, see Thompson (1998).

At the equator, warm moist air rises, cooling adiabatically as it does so (see pp. 159–60) and producing rain in the tropics. This draws in north-easterly winds in the

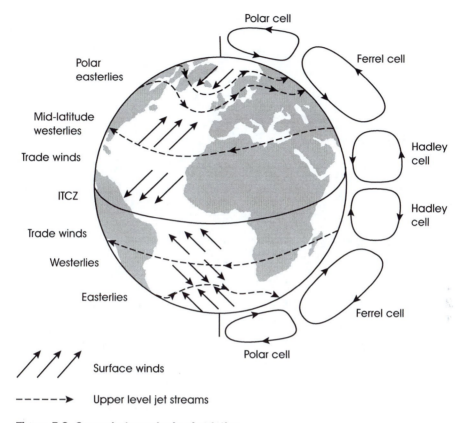

Figure 5.2 *General atmospheric circulation.*

Atlantic, north of the equator – the trade winds. As the upper airstream moves to the north at altitude it cools, and then falls back to ground level, now dry having lost all its moisture as rain. This produces the deserts of the Sahara, the Middle East and the south-western USA, together with their counterparts in the southern hemisphere – the Kalahari and central Australia. This cycle produces the first of three pairs of cells, known as the Hadley cells. In between the northern and southern trade winds lies the intertropical convergence zone (ITCZ) or the doldrums, a region of low winds and low rainfall. There is some seasonal variation, as the maximum solar input moves from the equator to each tropic and back, which moves the ITCZ and produces the wet and dry seasons of the monsoons of some tropical countries.

The next cells are the Ferrel cells, where circulation is in the opposite sense. Here the equatorial air that has descended produces the south-westerly winds of the mid-Atlantic, as it still has westerly momentum from its equatorial motion. These bring warm wet air to the western coast of Europe. The final, polar, cells produce cold north-easterly winds that circulate round the Arctic, and south-easterlies around the Antarctic.

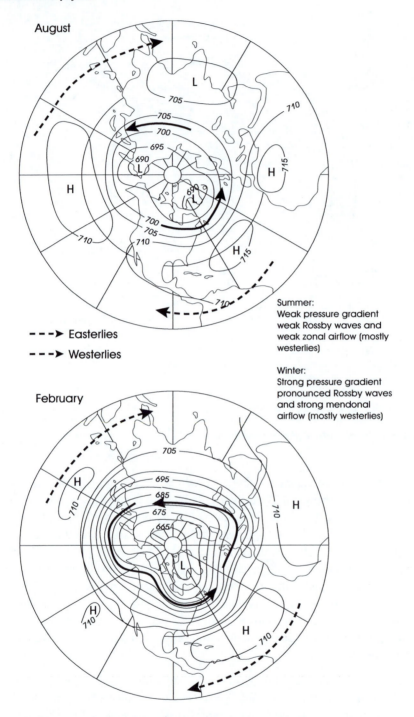

Figure 5.3 *Idealised airflow at 3 km elevation in summer and winter. Rossby waves are most pronounced in winter, with a higher pressure gradient.*

Source: reproduced with permission from Thompson (1998) p. 119

Surface level winds are accompanied by high level winds at the top of the troposphere into the stratosphere, which travel around the globe balancing out the mass transfer of air. These winds form the jet streams, narrow bands of high-speed winds reaching 160 km h^{-1} in speed. In mid-latitudes, as the wind moves to the north, it is subject to the Coriolis force (see p. 21), tending to divert it in a clockwise direction (in the northern hemisphere), swinging around under its own angular momentum until it is blowing in a southerly direction. As it moves away from the pole, the Coriolis force weakens because of the curvature of the Earth, and the airflow is subject to a force in the opposite sense due to its own vorticity, or the conservation of its angular momentum around the pole. This diverts it once more towards the pole. Together with momentum and pressure variations, the result is that these strong winds blow in a standing wave pattern around the pole, at between 30–60° of latitude. The waves are termed Rossby waves (Thompson 1998: 113), and can number between three and six around the Earth, with a wavelength of several thousand kilometres (Figure 5.3). Their position is also thought to be influenced by topography.

Weather disturbances

Many features of the weather we experience are due to secondary circulation, specifically the features of low- and high-pressure systems. These systems develop within the general circulation discussed above, as a result of the interaction of air masses of different temperature and motion, although the positions of land masses also play a part. Low pressure systems are generated where air is rising in the general circulation, along the ITCZ and the boundary around 50° latitude between polar and mid-latitude cells, where air at different temperature from equator and poles meets. High pressure systems arise where air is sinking, between the Hadley cell and the mid-latitude cell. At a more local level, tertiary weather disturbances include tornadoes (pp. 24–5), thunderstorms, land-sea breezes and mountain winds.

The physical processes that produce these disturbances are similar, whether they originate from frontal interactions between air bodies with different properties from the different cells, or as a result of differences between marine and land-based air masses. The key physical processes can be summarised as:

- convection from temperature differences due to: insolation; adiabatic temperature change with pressure; latent heat exchanges with condensation and evaporation; land and sea heat capacities; elevation and radiative cooling;
- the Coriolis force due to the Earth's rotation; and
- dynamic effects resulting from conservation of momentum and angular momentum.

Low-pressure systems form as warm air rises, either due to thermal or dynamic effects, cooling adiabatically as it does so, leaving a low pressure region at sea

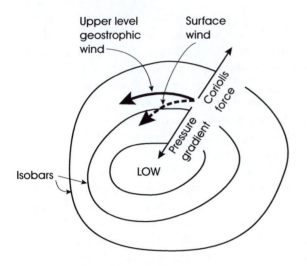

Upper level geostrophic wind

Surface wind

Coriolis force

Pressure gradient

Isobars

LOW

Figure 5.4 *Cyclonic weather system.*

level. As the air cools, water will condense out, producing clouds and releasing latent heat that can fuel storms. Where upper airflows reinforce this surface movement the low-pressure zone will be intensified, as the rising air is carried away at high level. Vorticity is created, with the direction of rotation dependent on the equilibrium between the force due to the pressure gradient, perpendicular to the isobars, which tends to pull the wind towards the centre of the low-pressure area; and the Coriolis force which diverts it outwards. At equilibrium these two forces will be equal but in opposite directions, resulting in a wind blowing perpendicular to both, i.e. parallel to the isobars, around the centre of low pressure, as shown in Figure 5.4. This is termed a **geostrophic** wind.

Upper level winds are most closely geostrophic, while at the surface winds will spiral slightly into the centre of the low pressure system. These are cyclonic systems (as distinct from tropical cyclones, Box 5.1), moving anti-clockwise in the northern hemisphere, clockwise in the south. Cold and warm fronts are associated with the low-pressure system as air masses of different properties (e.g. temperature, moisture content) meet. Strong winds and cold wet weather are associated with the lowest depressions, such as that common in northern Britain (Plate 5.1). Closer to the equator, tropical cyclones may form from low-pressure systems, from the heat of the sun.

The converse situation is a high-pressure system. Here cold air sinks, in Europe due to cold air subsidence from the poles or in the tropics due to dynamic flows. As it sinks it warms adiabatically, evaporating any water, leading to clear and cloudless conditions. The high pressure at the surface causes the air to tend to diverge away from the centre. Due to the Coriolis force the air spirals, down and outwards, in an anti-cyclonic system. As for the low-pressure system, it will be intensified only if upper airflows reinforce the movement, such as the jet stream supplying the air for the outwards motion.

Monsoon winds are seasonal airflows from the oceans on to the land in tropical areas including southeast Asia, India, northern Australia and Africa, producing large amounts of rain. The predominant mechanism is the seasonal heating of the land relative to the sea producing a pressure difference. This is propagated by changes in the upper airflow. In India a movement of the jet stream to the north of the Himalayas allows incursion of low-pressure systems further south on to the subcontinent, while elsewhere similar seasonal changes in upper airflows influence

Plate 5.1 *North Atlantic depression off northern Britain, 31 August 2000.*
Source: reproduced by courtesy of Dundee Satellite Receiving Station

the pressure differences that produce the monsoon. In the dry season the pattern is reversed, producing dry winds from land on to sea. A late or failed monsoon can be disastrous for agriculture, which may occur due to variations in the general circulation and upper airflows. One factor is the amount of snowfall in the Himalayas and in Tibet, while the El Niño circulation of the Pacific is also influential.

Clouds

Clouds form when moist air cools, most commonly as a result of adiabatic cooling when air rises and the pressure decreases. Air gains moisture from the sea or damp ground and rises, whether because of ground-level temperature increase, incursion

Box 5.1

Tropical cyclones

Tropical cyclones are powerful storms originating over oceans in the tropics, known in different parts of the world as hurricanes, cyclones or typhoons. In a tropical cyclone, energy is input from warm surface sea water as a result of evaporation followed by condensation. Once the low-pressure region has developed as described above, the storm fuels itself as it moves slowly across warm water (above 26°C), lifting warm moist air which, as it rises, releases its latent heat as water condenses out, increasing the uplift. This process produces torrential rain and high winds around a clear 'eye' of descending, clear air (Figure 5.5).

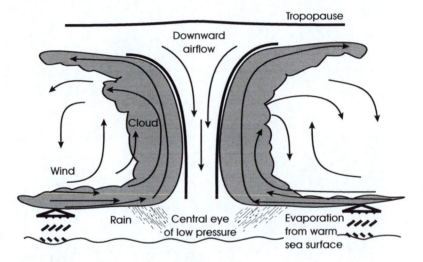

Figure 5.5 *Tropical cyclone – vertical cutaway.*
Source: adapted from Thompson (1998) p. 160

These highly destructive storms may be hundreds of kilometres across, with winds as high as 240 km h^{-1}, together with flooding and sea surges. They form in a 'hurricane belt' along the ITCZ where the trade winds converge, in late summer when this lies north of the equator. Within 5° of the equator they cannot form, as here the Coriolis force is not strong enough; while further north or south than around 15° of latitude, sea surface temperatures are not sufficient to maintain the necessary input of latent heat. Once formed they can travel quite large distances across warm oceans. Once the cyclone reaches land, it will quickly subside, as the latent heat is no longer available.

The hurricane that hit the south coast of England in October 1987 originated as a tropical cyclone in the Caribbean and tracked across the Atlantic, enabled by seasonally high sea temperatures, then up the European coast to reach Brittany and finally came inland in Sussex. It has been suggested that their frequency has increased this century and will increase further due to global warming, as a result of increases in sea surface temperature. They could also become more common at higher latitudes.

(a)

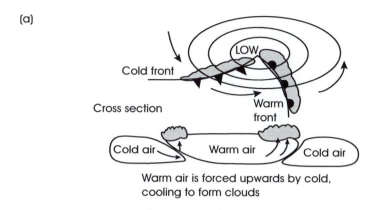

(b)

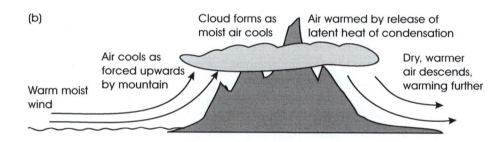

(c)

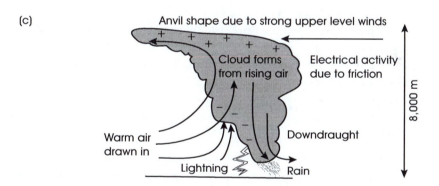

Figure 5.6 *Cloud formation: (a) in a low-pressure system; (b) due to mountains; (c) thundercloud.*

of colder air, low-pressure systems and upper airflow or the presence of mountains. At a certain height the air cools enough to reach the dew point and vapour starts to condense, producing a cloud base. When droplets become large enough they will form as rain, or as snow if they condensed below freezing point.

Where the prevailing wind is from the sea on to a mountainous land mass, the forced upward movement will produce clouds and rain on the mountain slopes. As heat is released by the condensation of the rain, when the air returns to ground level it will be warmer than before the rain occurred. The difference is due to the difference between the dry and wet adiabatic lapse rates (see p. 160). This effect produces the warmer drier climate of rainshadow areas.

Thunderclouds develop principally as a result of warm moist air rising in unstable atmospheric conditions. Thunderstorms commonly develop as a result of the Sun's heat in the afternoon or early evening in summer in the mid-latitudes and in the moist tropics. As the air rises the moisture condenses releasing its latent heat producing further convectional updraughts, which may reach between 4–20 km in height. The rapid air and cloud movements produce static electricity that forms lightning, while thunder results from the rapid expansion and contraction of air as the lightning discharges. Ground-level winds replenish the updraughts, and heavy rain ensues, with large drops due to rapid condensation in the strongly rising air.

Clouds have many types, according to the droplet size, height, moisture content and temperature which may result in ice rather than water droplets. The speed of formation and wind speed affect these types, along with other atmospheric conditions. Mist and fog are essentially ground-level clouds, with small droplets, forming in cold wet conditions, while in some mountainous areas the permanent cloud cover around the slopes can produce unique cloud forest ecosystems such as those of Central America. In order for water vapour to condense, it requires condensation nuclei (see p. 141), so the presence of dust or pollution can produce more clouds with smaller droplet sizes.

Clouds and rain will dissolve any soluble pollutants from the air, which affects atmospheric chemistry and may result in the deposition of the pollutant a long distance from its source. Long distance transport of pollutants is evident in the case of sulphur dioxide and acid precipitation, where some of the emissions from the UK are finally deposited in Scandinavia. These pollutants are also present in mists and fog, which can have serious environmental consequences as they can be absorbed by plants and animals directly, particularly in highland areas and cloud forests or in smog.

The radiative properties of clouds are vital to their influence on climate, but are highly complex. Water vapour, liquid water and ice all absorb radiation at a number of different wavelengths, some of which contribute to greenhouse warming. However clouds also reflect radiation at their surface, and scattering of shorter wavelengths occurs particularly in finer mists. It is a common observation that a clear night is cooler than a cloudy one as radiative cooling is greater, while in summer the hottest days are obviously the cloudless days. For these reasons the net effect of clouds on surface temperature is difficult to estimate with any certainty, and could be either positive or negative. The effect at any one location depends upon the cloud type, height, droplet size and crucially whether water or ice, and diurnal or seasonal variations.

Ocean currents

Ocean currents play an important role in transferring heat around the globe, as water has a high heat capacity compared with air, although the currents are far slower than winds. Currents are also crucial in dissolving CO_2 from the atmosphere and carrying it to a sink (see p. 208), in the form of dissolved CO_2 at depth or fixed biologically by phytoplankton. To some extent the main ocean circulation is similar to the general circulation of the atmosphere, together with eddies that correspond to secondary weather systems. Currents are driven by surface winds, differences in temperature and salinity which affect density, and the shape of ocean basins. The major features are summarised here, and are described in more detail by Summerhayes and Thorpe (1996).

The ocean is differentiated by layers of different temperature, and hence different densities. The surface 100 m or so is generally warm from the Sun's heat and well mixed, due to wave action. It is within this layer that atmospheric gases are dissolved, and phytoplankton lives. Below this level comes the thermocline, a region where temperature declines down to about 1,000 m depth. The remaining deep water, which comprises 80 per cent of ocean volume, is constant in temperature at around 4°C (the temperature at which water is most dense). These layers can tilt, due to pressure and density changes, causing them to flow. Ocean water also differs in salinity, due to fresh meltwater from ice, salt concentration as ice freezes, incoming freshwater, and evaporation, which affects density. These density variations, as a result of both temperature and salinity, give rise to pressure variations, which produce currents. Because of the Coriolis force these currents are geostrophic, running parallel to the pressure gradient in the same way that winds blow along the isobars.

Surface winds tend to pile up surface water, pushing it towards one side of the ocean basin, such that sea level can vary by up to a metre. This piling up of warm surface water in turn distorts the level of the thermocline, tending to push deep cold water back the other way, and promoting upwelling of colder water at certain sites.

The result of these forces is a 3-D circulation pattern throughout the world's oceans. At the surface, circulating currents called gyres form, at both subpolar and subtropical latitudes, in which warm currents flow towards the poles at the western edge of the oceans, and cold water towards the equator along their eastern edges. The North Atlantic Drift forms part of one of these gyres. A separate current system operates around the equator where the Coriolis force does not have an effect. There is also the Antarctic Circumpolar current around the globe in the uninterrupted Southern Ocean, which connects all the other ocean basins.

The surface currents are part of a larger picture, as water is being drawn down at a small number of sites in the Arctic and Antarctic and flowing towards the tropics at depth. Deep convection, the process drawing cold water down to the ocean depths, occurs at the edge of the Greenland ice sheet and in the Weddel Sea near

Box 5.2

El Niño

'El Niño' is a periodic current system in the Pacific that normally flows once every few years, sometimes referred to together with the normal current system as El Niño/Southern Oscillation or ENSO. It has become of great interest to climatologists because, in the last twenty years or so, it has become more frequent than ever before. In an El Niño year, climate across the southern hemisphere is altered dramatically, bringing higher temperatures, droughts or floods and abnormal wind patterns, not just to Pacific areas such as Australia and Indonesia but as far away as India and Africa.

In a normal year, the principal wind and current system in the equatorial Pacific carries warm water away from the coast of South America in a westerly direction, producing a warm pool in the west around Indonesia. Cold water upwells off the south American coast to supply this current. Every three to seven years, the system is reversed giving rise to an El Niño year, when the warm water flows back across the Pacific towards south America (see Figure 5.7). This brings with it heavy rain to soak South Pacific islands and the coast of Peru, leaving droughts in Indonesia and Australia as monsoon rains fail. As the cold water brings with it nutrients that form the basis of the food chain for fish, an El Niño event also reduces fishing yields in south America.

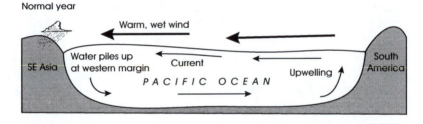

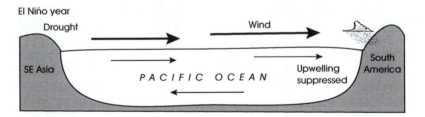

Figure 5.7 *El Niño.*

It has been suggested that the recent unusually high frequency of El Niño years is due to global warming (see for instance Pearce 1999). Since the mid-1970s sea surface temperatures have been increasing and the thermocline – the boundary between warm surface water and colder deep water – has been getting deeper. Warmer sea temperatures promote the El Niño current, and the upwelling of cold water is not as strong, suppressing the normal current system.

Antarctica. This process is driven by changes in density as sea ice forms, freezing out fresh water and leaving cold water with higher salt content that is therefore denser and sinks. Because it is driven by both temperature and salt content it is termed **thermohaline circulation**. The pumping action forces the deep sea water slowly towards the tropics, where it will gradually rise to the surface. This upwelling of cold water carries nutrients to warm surface waters, replenishing plankton levels and greatly increasing productivity of these marine ecosystems. It also serves to maintain the stability of the thermocline. The complete system is sometimes referred to as the oceanic conveyor belt. The process is very slow, with complete circulation taking hundreds of years before the water returns to the surface. In studying climate change it is of foremost importance, as it controls the rates at which CO_2 is absorbed and at which the deep oceans are warming, as well as being ecologically important.

It is possible that relatively small changes in temperature and radiation could bring about major changes in currents, leading to large and chaotic (i.e. unpredictable) changes in climate. For instance, it has been suggested that as higher temperatures reduce the formation of sea ice around Greenland, the deep convection that drives ocean currents in the Atlantic could be switched off, resulting in reduced flow in the North Atlantic Drift and far lower temperatures across much of north-western Europe.

Microclimates

The term microclimate refers to a specific set of climatic conditions affecting a small area. This may be a forest or other natural ecosystem, an urban area (see Box 5.3), or a topographical feature such as a valley. Many factors affect microclimate, principally topography; shelter from wind by landform, buildings or vegetation; availability of moisture and heat capacity of the surface. Microclimates can provide conditions for specific plants and support distinctive ecosystems as a result, which in many cases provide feedbacks and help maintain the microclimate. The classic example is a rainforest, which cycles water through evapotranspiration, increasing atmospheric water and increasing rainfall. Deforestation can reduce the rainfall resulting in arid, unproductive land that is unsuitable for farmland and vulnerable to erosion.

Winds at the surface are affected by friction with the surface of the Earth, reducing their speed. The increase of windspeed with height depends on the surface roughness, which is lowest for clear water or flat sand, and higher for grassland, higher still for forests or cities. As the lower wind speed reduces the Coriolis force, this also alters the direction of the wind, to be angled more directly towards low pressure at an angle of 10°–30° to the isobars. This results in winds following a spiral pattern with height, termed the Ekman spiral after its discoverer. Layers of wind are increasingly deflected until at heights of around 500–1,000 m the surface

Box 5.3

Sweltering cities

Cities are generally warmer than surrounding rural areas, due principally to the non-reflective nature of roads, concrete and building materials, and their high heat capacity. During the day, more of the Sun's radiation is absorbed by these dark surfaces, heating them up. At night, the surfaces re-radiate the heat, warming the air above the city. The result is air temperatures elevated by typically 3°C, but up to 7°C in some cases. Surface temperatures can be far higher. In one study in Atlanta, Georgia (Anon 2000), city rooftop temperatures reached 70°C, sufficient to regularly create its own thunderstorms. During the 1996 Olympic summer, in one nine-day period there were five days when Atlanta's own heat triggered thunderstorms.

The heat island effect is aided by waste heat from energy used in cities (cars, building heating systems or air-conditioning, electrical appliances and so on) although this effect is

Plate 5.2 *Surface temperatures around the English Channel. This is a thermal (12 μm) image of the English Channel by night, taken on 7 September 1991, where London and other urban areas can clearly be seen because of their higher temperature. Land temperatures range from 5–15°C, while sea temperatures are higher than land, with coastal water reaching 17°C. Other features such as clouds and the River Seine are also visible due to their temperatures.*
Source: reproduced by courtesy of Rutherford Appleton Laboratory

relatively minor. Smog and shelter from wind can also increase temperatures, while humidity is generally lowered by the lack of vegetation and open ground.

Across the world, urbanisation is occurring at an increasing rate, increasing this problem particularly in tropical areas. It has been suggested that measures such as tree-planting and painting roofs white will need to be taken, to reflect more heat back into space and alleviate the temperature increases.

no longer has an effect and the Coriolis effect dominates. Here winds will be geostrophic, once again running parallel to the isobars around the centre of low pressure.

The ozone layer

Ozone in the stratosphere acts to protect us from harmful ultraviolet (UV) radiation from the Sun, which it absorbs. A small reduction in ozone concentrations can lead to large increases in the amount of harmful UV reaching Earth, specifically around wavelengths of 295–300 nm. Excessive UV exposure can cause skin cancers, and in large amounts can also be harmful to plant growth, as discussed on pp. 127–8. The highest risks are in spring at extreme southerly latitudes or to a lesser extent at extreme northerly latitudes, corresponding to the formation of the ozone holes over Antarctic and Arctic polar regions.

The stratospheric ozone layer should not be confused with tropospheric, or low-level, ozone pollution. Ozone forming near ground level from vehicle emissions and in photochemical smog is a potent oxidising agent and damaging to health. The ozone layer is also a separate issue from climate change and greenhouse warming, although there are some connections. The CFCs that cause ozone depletion are greenhouse gases, but destruction of the ozone layer reduces the greenhouse effect slightly as the ozone also acts as a greenhouse gas. Increasing CO_2 can increase ozone depletion by a mechanism that will be discussed later in this section. Conversely if ozone destruction were to become serious enough to significantly reduce primary productivity of phytoplankton and forest ecosystems, as a result of UV damage, this would affect climate change in reducing these important sinks for atmospheric CO_2. As with many environmental systems, all things are inter-connected and a holistic view is necessary.

Ozone (O_3) forms from oxygen (O_2) by a reaction under the influence of UV radiation. Short wavelength radiation, of wavelengths 175 nm or less, has sufficient energy to dissociate oxygen into two separate free oxygen atoms. This occurs at heights of 50 km or more, and results in very little radiation less than this wavelength penetrating further. The free O atom can then combine with an O_2 molecule to form ozone. Ozone forms in this way above tropical and equatorial regions where solar radiation is highest, and spreads around the globe to form a thin layer at between 20 and 26 km in height. The ozone layer if expressed at standard temperature and pressure would only be 3 mm thick, yet it absorbs strongly enough in the UV region to act as an almost complete shield against radiation of wavelengths less than around 295 nm.

$$O_2 \rightarrow 2O \qquad \text{Oxygen dissociates in presence of UV radiation}$$

$$O_2 + O \rightarrow O_3 \qquad \text{Ozone formation}$$

However in the presence of certain pollutants, chiefly chlorofluorocarbons (CFCs), ozone can be destroyed. Free chlorine (Cl) radicals act to strip O_3 of an oxygen atom, forming O_2 and ClO, which then loses the O to a free oxygen atom to form O_2 and Cl again (Boeker and van Grondelle 1999):

$$O_3 + Cl \rightarrow O_2 + ClO \quad \text{Ozone destruction by Cl radical}$$

$$ClO + O \rightarrow Cl + O_2 \quad \text{Cl radical free to act again}$$

The net effect is to transform ozone into O_2, while leaving the Cl radical free to promote another reaction, thus one CFC molecule can destroy a large number of ozone molecules. Nitric oxide (NO) and hydroxyl (OH) radicals can act in the same way. The resulting reduction in ozone concentration depends upon a balance between the processes of ozone formation and destruction. The series of reactions that destroy ozone take place most readily at very low temperatures, and so occur mainly over the poles in early spring. In the stratosphere, unlike the troposphere, there is very little water vapour and so generally no clouds. During the Antarctic winter temperatures plummet, and when the temperature of the stratosphere drops below around $-80°C$, polar stratospheric clouds can form. It is these clouds that allow the formation of highly reactive free Cl radicals. In early spring as the first rays of sun strike the clouds the reactions that destroy ozone commence. This results in the so-called ozone hole over the Antarctic, which can be measured by means of remote sensing of atmospheric absorption. The Antarctic ozone hole has occurred every winter for the past two decades with increasing severity. During the summer the ozone is replenished from the tropics. Ozone loss in the Arctic is generally less severe simply because stratospheric temperatures are not usually low enough to allow clouds to form, although in recent years an Arctic ozone hole has also formed, reaching a record low in 2000 (see Walker 2000).

While production of the most harmful CFCs is now subject to strict international controls stemming from the Montreal Protocol, they are very long-lived once they reach the stratosphere, and are still being emitted from many sources. The ozone holes appear recently to have deteriorated more rapidly than hoped or expected, particularly in the Arctic. It is thought this may be due to a feedback with climate change. Increased CO_2 levels in the troposphere hold the Earth's heat at the surface, resulting in lowered temperatures in the stratosphere. These lower temperatures promote the formation of the polar stratospheric clouds that allow ozone-destroying reactions to occur more rapidly. Hence the ozone holes have increased more rapidly in both size and level of depletion.

Climate change

The Earth's climate is constantly changing, as in the past it has been subject to changes due to volcanism, solar activity, variations in its orbit and meteorite impacts, together with the evolution of life. The development of the atmosphere has acted to maintain temperatures at fairly constant, equable levels conducive for life to survive. This is due to the natural greenhouse effect, holding heat at the surface of the Earth. Without the atmosphere, the Earth's surface would fry by day and freeze by night. However the composition of the atmosphere has not remained constant. Over billions of years, photosynthesis by phytoplankton and plants has acted to remove carbon dioxide from the atmosphere and replace it with oxygen, tying up the carbon in the form of carbonaceous rocks such as limestone and chalk, and as fossil fuels. This has created the current climate in which today's land-based life-forms evolved. Over more recent millennia, the climate has been subject to periodic ice ages and interglacial warming, with variations of a few degrees in temperature causing quite large changes to ecosystems.

Anthropogenic climate change or global warming centres around the fact that carbon dioxide from fossil fuels, along with other greenhouse gases, is being released into the atmosphere. These greenhouse gases absorb heat radiation and increase surface temperatures, having potentially large and irreversible effects on climate.

The Earth's radiative balance, albedo and the 'greenhouse effect'

Like any warm object, the Earth emits heat radiation. It also absorbs radiation from the Sun. It is a balance between these two heat fluxes that determines the surface temperature of the Earth. The Sun's radiation strikes the atmosphere, where some is absorbed and some reflected. The rest, mainly at visible wavelengths, is transmitted to the Earth's surface, where some is reflected and the remainder is absorbed to heat up the ground. The warm ground then radiates at infra-red wavelengths back through the atmosphere, where some is absorbed and some transmitted out into outer space. This is shown in Figure 5.8.

On average, the Earth receives 340 W m^{-2} of energy from the Sun, measured at the top of the atmosphere. Of this, 100 W m^{-2} is reflected and 240 W m^{-2} is absorbed by the Earth and its atmosphere. To retain the same temperature, this means the heat being radiated out must also be 240 W m^{-2}, measured at the top of the atmosphere. This is the radiative balance for the top of the atmosphere It could also be measured for the Earth's surface or between layers of the atmosphere: at constant temperature, incoming radiation = outgoing radiation.

An increase in greenhouse gases such as CO_2 increases the amount of infra-red absorbed by the atmosphere. This radiation is then re-emitted, warming the Earth's surface and lower atmosphere. This is known as **radiative forcing**. The equilibrium

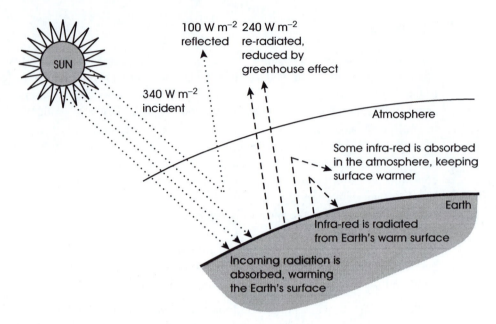

Figure 5.8 *The Earth's radiative balance and the greenhouse effect.*

between incoming and outgoing radiation is upset, producing a permanent temperature change to redress the balance (Box 5.4). This mechanism is present in the warming of a greenhouse, as glass and CO_2 both absorb infra-red wavelengths but allow visible light to pass through. However in a real greenhouse, shelter from wind and reduced losses by convection are more important in the increased temperature attained, which are not relevant in the atmosphere, so the term 'atmosphere effect' is more accurate.

The amount of energy reflected from the Earth's surface depends on its **albedo**. The albedo is the proportion of energy reflected, shown in Table 5.1. Hence for snow, up to 95 per cent of the Sun's energy is reflected, with only 5 per cent absorbed to warm it, while for dark asphalt 5 per cent is reflected and 95 per cent absorbed.

If albedo is high, more energy is reflected, leading to a cooling effect. As ice has a high albedo, this maintains lower temperatures at the poles and other large areas of snow and ice, and helps prevent ice from melting. Conversely, low albedo areas – including roads and cities (see Box 5.3) – absorb more radiation, and so become warmer. Albedo is an important factor in both microclimates and climate change as it can be altered by many anthropogenic factors.

The amount of radiation absorbed by a gas is described by its absorption spectrum (p. 118), which shows how much radiation is absorbed by the gas over a range of

Table 5.1 *Albedo of various surfaces*

Surface	Albedo
Fresh snow	0.8–0.95
Dry sand	0.4
Leafy crops or forests	0.2
Calm sea water	0.05
Dark asphalt	0.05

Box 5.4

Global warming on the back of an envelope

It is estimated that the effect of a doubling in concentration of greenhouse gases such as CO_2 will reduce the heat being radiated out, at the top of the atmosphere, from 240 W m^{-2} to 236 W m^{-2} at current temperatures. It will disrupt the radiative balance, and cause an increase in temperature to restore equilibrium. From this figure it is possible to estimate very simply the increase in temperature resulting from the greenhouse effect using Stefan's Law (Equation (3.19)):

$$E = \varepsilon\sigma T^4$$

where E is energy emitted at temperature T from a body with ε the emissivity and σ is Stefan's constant. Thus if T_1 and T_2 are the temperatures before and after the change, and T_1 is 300 K:

$$\varepsilon\sigma = E_1/T_1^4 = E_2/T_2^4$$
$$T_2 = \sqrt[4]{(E_2 T_1^4/E_1)}$$
$$= \sqrt[4]{(240 \times 300^4/236)}$$
$$= 301.26 \text{ K}$$

In other words, doubling CO_2 concentration produces a warming of 1.26 degrees. This really is a back of the envelope calculation, as it makes certain hidden assumptions about heat transfer through the atmosphere, and fails to take into account the many complex feedback effects on Earth's climate that may either mitigate or worsen the actual effect at the surface. However it does give an order of magnitude estimate that is surprisingly close to that predicted by highly complex climate models. This small change is sufficient to have a big impact on climate.

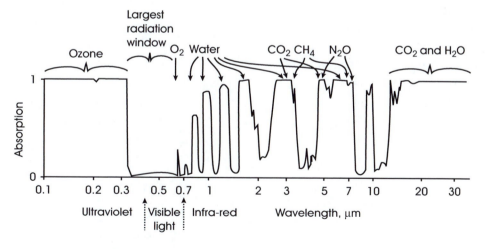

Figure 5.9 *Atmospheric absorption spectrum.*
Source: adapted from Fleagle and Businger (1980) p. 232

wavelengths. For the atmosphere, an absorption spectrum can be measured depending on its constituent gases, shown in Figure 5.9. Regions where the atmosphere is transparent are known as **radiation windows** – little radiation is absorbed at these wavelengths. The largest radiation window is around the visible spectrum (0.4 to 0.7 μm). Greenhouse gases have absorption spectra that fill windows in the atmosphere (in the IR wavelengths), where otherwise radiation could pass through.

Greenhouse gases and greenhouse warming potentials

The most important greenhouse gas that we emit is carbon dioxide (CO_2), but there are a number of others, as shown in Table 5.2.

For greenhouse gases other than CO_2, a greenhouse warming potential (GWP) is defined. This is used to convert the concentration into 'CO_2-equivalents', i.e. the amount of CO_2 emitted that would have the same effect on warming as 1 tonne of methane, CFCs etc. Because the gases have different lifetimes in the atmosphere, their greenhouse warming potentials depend on the timespan over which they are measured.

Note that CO_2 is a relatively weak greenhouse gas, but is the most important because it is emitted in such vast quantities. By contrast CFCs are thousands of times stronger in their warming effect, but they are not emitted in such high quantities.

Table 5.2 *Properties of the main greenhouse gases*

	Carbon dioxide	Methane	CFCs and HCFCs	Nitrous oxide
Contribution to GH warming	55%	15%	24%	6%
Concentration:				
Preindustrial	280 ppm	0.8 ppm	0	288 ppb
Current	359 ppm	1.7 ppm	800 ppt	310 ppb
Increase p.a.	0.5%	0.9%	4%	0.25%
Lifetime, years	50–200	10	50–150	150
GWP: 20 y	1	63	4,000–7,000	270
100 y	1	21	1,500–7,300	290
Sources	Fossil fuels, deforestation, cement making	Agriculture, gas leaks, termites	Propellants, refrigerants, fire extinguishers, agricultural chemicals	Combustion

Note: ppm = parts per million; ppb = parts per billion; ppt = parts per trillion, all by volume.

CO_2 absorbs radiation with wavelengths of between 13 μm and 100 μm. Because CO_2 is naturally present in the atmosphere, relatively little radiation in this range escapes, so the marginal effect of increasing CO_2 is small. The other greenhouse gases however absorb at shorter wavelengths, between 7 μm and 13 μm, which is an important atmospheric 'window' – the natural atmosphere absorbs very weakly in this region. The marginal effect of small increases in these gases is therefore much larger than for CO_2 and so their greenhouse warming potentials are much higher.

Another important greenhouse gas is water vapour, which overall has a greater impact than any. This is not included in Table 5.2 because it is so variable – the concentration depends chiefly on temperature, humidity varies widely, and clouds also have an effect. While there are no emissions as such, water vapour may increase as a result of climate change and is very important in many feedback mechanisms discussed later.

Greenhouse warming, feedbacks and climate impacts

In order to estimate the impacts of global warming, it is necessary to follow the effects of a greenhouse gas through from its emission to the final climatic changes, as illustrated in Figure 5.10.

The first step is to estimate emissions: this is relatively straightforward for CO_2 from energy use, but more difficult for some of the other sources and greenhouse gases.

Once emissions have been quantified, the impact on atmospheric concentration is determined from knowledge of the cycling of the gas in the biosphere. This is not

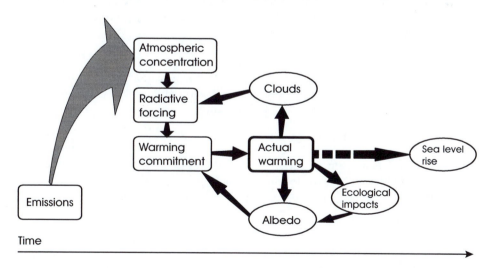

Figure 5.10 *Stages in greenhouse warming, showing some of the main feedback effects.*

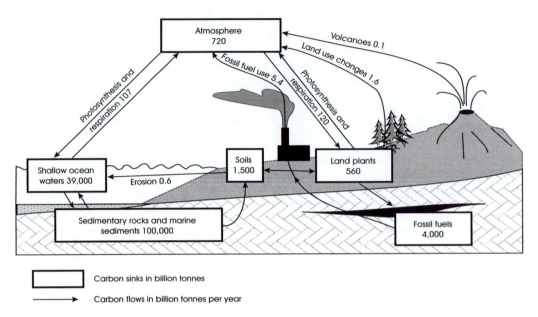

Figure 5.11 *The carbon cycle.*

Source: adapted from Botkin and Keller (2000) p. 62

known accurately, but can be estimated from historical records of emissions and concentrations. For carbon dioxide, approximately half of all emissions are rapidly taken up into sinks such as forests and plankton, with the rest increasing the atmospheric concentration. Over a long time period, (around 200 years), if CO_2 emissions stopped altogether, these sinks would gradually absorb all the excess CO_2 and atmospheric concentration would stabilise at something close to its pre-industrial level. These changes depend on the **carbon cycle**, an important biogeochemical cycle that describes the flows and sinks of carbon, determining how much is in the atmosphere. Some of the major known flows and sinks are illustrated in Figure 5.11.

The change in the atmospheric absorption spectrum as a result of emissions, and how much IR radiation is blocked, is described by radiative forcing. The change in radiative forcing is given approximately by:

$$\Delta F = 6.3 \log_e (C/C_0) \tag{5.1}$$

where C is the concentration now, and C_0 is the pre-industrial concentration. The effect on radiative forcing of doubling of atmospheric CO_2 concentration can be estimated from Equation (5.1):

$$\begin{aligned}
\Delta F &= 6.3 \log_e (C/C_0) \\
&= 6.3 \log_e 2 \\
&= 4.4 \text{ W m}^{-2}
\end{aligned}$$

In other words, as a result of CO_2 doubling, 4.4 W less energy would escape from every square metre of the Earth than is incident upon it, and the equilibrium temperature would rise to redress the balance. The log relationship indicates a 'diminishing return' to additional concentration. Doubling the concentration from pre-industrial level to 540 ppm would increase radiative forcing by 4.4 W m^{-2}; a further doubling to 1,080 ppm would be needed to increase by another 4.4 W m^{-2} and so on. The reason for these 'diminishing returns' is that, as CO_2 concentration in the atmosphere increases, it blocks most of the IR in its absorption spectrum. So for each additional increase, there is less IR escaping to be blocked. At higher concentrations, increasing CO_2 further only affects the small amount of IR still being let through, having less and less impact. This relationship is known quite accurately, from both theoretical and experimental results.

Other greenhouse gases have different relationships to radiative forcing. For CFCs, because the natural atmospheric concentration is zero and the current level is low, the increase in radiative forcing is much higher. This is because these gases are filling the natural windows in the absorption spectrum – the diminishing returns do not set in until the windows start to be significantly blocked.

Once radiative forcing has been estimated, a warming commitment is defined as the long-term change in **equilibrium temperature** that this radiative forcing will bring about. The change in radiative balance induced by greenhouse gases reduces the amount of heat that is radiated out into space. To regain equilibrium between incoming and outgoing radiation, the Earth will warm up at the surface. From this effect alone, doubling CO_2 would result in an equilibrium temperature rise of about 1.3°C (see Box 5.4). After a certain time lag (several decades), the actual temperature change will gradually occur. The lag is due to the time taken for the new equilibrium to re-establish, as land and water masses warm up. The time lag for sea level rise is much longer – once the surface temperature increases, the temperature of lower layers of the oceans and the size of continental glaciers will gradually change leading to a new, higher, equilibrium sea level only after around 200 years. So we are now witnessing the warming and sea level rise due to the last 200 years of emissions.

Expected warming commitments and actual temperature changes are calculated by global models, as discussed on pp. 213–8. These models have a good correspondence with historical trends, but need to be highly complex to include the many factors influencing climate. Predicting changes such as regional variations, precipitation changes and ecological impacts is even more difficult, and the results always represent statistical trends subject to annual variations – the actual temperature in any one year can never be forecast with any certainty.

The picture is made more complex by the action of feedback mechanisms. First, increasing surface temperature will increase water vapour in the atmosphere. As water vapour is also a greenhouse gas, this will trap more heat, and increase the overall temperature rise: a positive feedback. Including this effect increases temperature rise to about 1.9°C. However the increased moisture will tend to reduce the adiabatic lapse rate, as the wet adiabatic lapse rate is lower than the dry. This is because of the release of latent heat as the moisture condenses at a height. This effect mitigates global warming at the surface as it tends to warm the atmosphere higher up (Boeker and van Grondelle 1999).

Second, increasing temperature will reduce the areas of ice, both at the poles and in mountains. Ice has a very high albedo – it reflects most of the Sun's radiation back out again, keeping itself cool. Once it has melted, more radiation will be absorbed (by sea, rocks, plants or whatever replaces it). This will increase the temperature further: another positive feedback. Including this effect increases temperature rise to about 2.9°C.

Third, there will be effects because of changing cloud patterns. These are more complex, as the radiation absorbed depends on the cloud type, which may alter (as discussed on pp. 193–6). In some cases increased cloud cover will cool the surface, while in others clouds act to retain more heat or the structure of clouds will be changed leading to heating; in other areas higher temperatures will reduce cloud cover. It is not clear if the net effect is positive or negative feedback, and there may be significant local variations.

A further area of uncertainty is biological feedbacks, affecting the carbon cycle. Increasing CO_2 in the atmosphere can stimulate plant growth by the 'carbon fertilisation' effect, as the photosynthetic rate is faster when more CO_2 is available. This tends to reduce CO_2 in the atmosphere, reducing the warming rate, but is limited as it only affects some plants, and growth is also constrained by water or nutrient availability. Warmer temperatures and higher precipitation could also have this effect, producing negative feedback. However higher temperatures combined with drought in some regions could reduce plant growth and even kill entire forests, in particular natural ecosystems that do not have time to adapt. One study by the UK Hadley Centre for Climate Prediction even suggests serious dieback in the Amazon, due to reduced rainfall (Adler 2000). This would lead to a big positive feedback as the carbon held in this important sink is released. Similarly any changes in phytoplankton activity brought about by changed sea surface temperatures would have major biofeedback effects. Any changes in vegetation can create further feedback through albedo changes. As the albedo of snow is so high, deforestation in Arctic areas would produce a strong cooling effect if dark, absorbing coniferous forests are replaced by snowfields. Increased sea temperatures will also reduce the amount of CO_2 dissolved in the oceans, while any changes in ocean circulation could also reduce this CO_2 sink.

Overall, including all these uncertainties, warming of 1–3.5°C is expected from a doubling of CO_2, expected to occur around 2030. Increases in temperature are likely to be associated with other changes, in precipitation and the incidence of storms. In general terms, higher temperature gradients in the atmosphere increase energy input into the climate system, leading to stronger winds and more storms.

Hence an extreme wind or flood that previously would have been expected to occur once every fifty years may in the future occur once every twenty or ten years, with obvious consequences for engineering design. Higher temperatures will increase atmospheric water content as surface evaporation increases, which also increases the transfer of latent heat into the atmosphere. This need not necessarily lead to greater cloud formation or higher precipitation everywhere however, as higher temperatures will also reduce condensation in the atmosphere, and may produce drought as soil moisture will be reduced by evaporation. It is expected however that rain may generally become more intense. The actual changes will depend on the exact position of flows in the general circulation, and so regional impacts are very difficult to predict.

Ice ages and colder climates?

Many pollutants act to cool the climate rather than to warm it, by blocking the Sun's radiation as it is incident on the atmosphere. Dust, particulates in smoke, and sulphur dioxide aerosols all act in this way, by scattering light back up through the atmosphere before it reaches the surface. The concern over the 'nuclear winter' predicted that a nuclear holocaust would be followed by many years of extremely cold weather, preventing resumption of normal agriculture, due to the large amounts of dust and smoke in the atmosphere.

Cooling from pollutants has obscured the climate record, as they have been emitted by industry in large amounts simultaneously to CO_2 over the past hundred years or so. In the short term, the cooling effect tends to dominate, which has been an important factor in preventing the warming effect of CO_2 being statistically measurable. In the longer term however, as other pollutants are washed out of the atmosphere, their concentration no longer increases and the underlying warming trend due to CO_2 becomes more apparent, as is now the case. Inclusion of the cooling effect from other pollutants, in particular sulphate aerosols, does however reduce global warming and slow the rate of change of climate predicted.

There are also natural sources of these cooling gases, such as volcanic action and smoke from forest fires. Major volcanic eruptions have a marked effect on temperature due to dust and sulphur dioxide produced, some of which may reach high into the atmosphere taking some years before it is fully washed out. The effect of recent eruptions including Mount St Helens and Mount Pinatubo has resulted in one or two cooler years afterwards. Prehistorically, much larger volcanic events could have had more dramatic consequences, causing major ecological collapse and mass extinctions. Dust in the atmosphere following the impact of meteorites or asteroids has a similar effect – it is thought to be climate cooling following a major meteorite impact that brought about the extinction of the dinosaurs.

The Earth's climate history has included a number of ice ages, each lasting several million years and including a number of advances and retreats of the ice, which at its maximum covered a third of the planet's surface. The warmer periods between cycles of advancing ice are known as interglacials, and may last thousands of years. During the coldest periods, global mean temperature would have been between 5–10°C cooler. As the water for the ice came from the sea, as a result of snowfall over a long period, sea level would have been much lower, by about 130 m. The last ice age commenced around 2 million years ago and finished 10,000 years ago, although it is possible that the current climate represents a short interglacial period and glaciers will advance once more in the relatively near future.

There are several possible contributory factors to the commencement of ice ages. Cyclical changes in the Earth–Sun distance and in the tilt and eccentricity of the Earth's orbit (known as Milankovitch changes) are widely accepted as the main cause. Changes in atmospheric dust from meteorites or large-scale volcanic activity could also have cooled the climate. Another possibility is that plate tectonic activity, as the continents rearranged themselves over millions of years, could have disrupted the global circulation. This could have resulted in reductions in the heat distributed from equator to poles, resulting in greater ice formation at high latitudes (Montgomery 1995). Once ice sheets start to expand, the much greater albedo of ice reduces temperatures further. Thus a glaciated world could be a stable climate situation, as the ice reflects enough solar radiation to keep the Earth cool.

It is thought that at the commencement of an ice age climate change would be relatively slow, as it would take thousands of years for snow to build up into glaciers to cover the large areas. This allows a certain length of time for ecosystems to adapt or to retreat to a more favourable location. However there is some difference of opinion on how long this adaptation time was, as some studies suggest that short-term changes could trigger an ice age. As the ice retreated, temperature rises could have been rapid – possibly as fast as temperatures are rising today under anthropogenic influences. It is possible that a relatively small change in some external factor, in atmospheric conditions or in ocean circulation, could have caused the climate to flip into another stable state at a higher or lower temperature, resulting in changes of several degrees within decades rather than thousands of years. This is an important issue in the context of modern warming, as the length of time it takes climate to change determines both the success of ecosystems in adapting and the human costs. Changing agricultural systems, building flood defences or changing the built environment to suit a new climate takes a long time, and costs would be much lower overall if changes were gradual in nature.

At the end of the ice age, ecosystems moved towards the poles or up the mountains. One outcome of this movement is that geographically isolated ecosystems may be very similar. For instance in the Smoky Mountains of North Carolina, many species are found at high altitudes that are common to lowland Canadian forests thousands of miles further north. The cold forests that covered the Carolinas during the ice age retreated up the mountains and northwards, leaving this island of isolated, cold climate species in the mountains of the south.

Sea level

Sea level has also not been constant historically, subject to variations due to climate changes and plate tectonic shifts. As stated above, during the ice age the sea was as much as 130 m lower than today, as so much water was bound up as ice. Sea level in many areas is changing because of the movements in the Earth's crust – for instance south-east England is tending to get lower while the north and west of the country are lifting. Thus sea level is rising in the south-east but falling in the north and west. Some of these movements are due to isostasy, the elastic rebound of the lithosphere since the ice age (see p. 149).

The mean sea level over the Earth has been rising by 1–2 mm per year over the last few decades. In relation to climate change, sea level rises for two main reasons: first, thermal expansion of the oceans as they get warmer, and second, melting of ice caps. Both effects include long lag times, as heat takes a very long time to percolate down into the oceans and ice sheets.

The most important contribution is from thermal expansion. Mountain glaciers and small ice caps are also important, which in most areas of the world have been observed to be retreating already. The large Greenland ice sheet would also retreat and have an impact. Antarctic ice is unlikely to melt significantly, and in fact increased snowfall could increase its thickness. The Arctic is not so important as it is not on land, although there will still be albedo feedbacks from reduced ice cover.

The expected rise in sea level is about 50 cm by 2100. Because of the long lags between emissions, warming and then sea level rise, there is already a sea level rise commitment – a rise which could not be avoided even if all emissions ceased today. The lowest conceivable emissions scenario would still result in a rise of 15 cm by 2100.

These rises will have serious consequences in many countries as large populations live in low-lying, flood prone areas that will require further flood defences. One country that will be very hard hit is Kiribati, an archipelago of low-lying atolls in the South Pacific. Already, Kiribati is facing increased problems of coastal erosion and salination of its water supplies as the higher salt water table invades the groundwater. In the long run the nation may effectively cease to exist, as most of its land area will be underwater and its population forced into exile as environmental refugees.

Climate modelling – predicting change

Forecasts of climate change are found using global circulation models (GCMs). These large computer models divide the Earth's surface into grid squares and calculate a radiation balance for each. Surface parameters include temperature, albedo and topography. The atmosphere is divided into boxes, in three dimensions, so the model can calculate heat and momentum exchange between neighbouring boxes. This allows calculation of the pressure and hence winds, temperature, moisture content and precipitation for each box. Existing models can forecast the existing climate from a 'cold start' with reasonable accuracy – that is they can reproduce the basic general circulation system, temperatures and major weather features from their fundamental theoretical inputs and assumptions. They can then be used to look at increased CO_2 concentration (usually doubled). The expected change in global mean temperature is thought to be fairly accurate, but regional variations and changes in precipitation or wind/currents are less well known. As the computing power required is so vast, the scale of resolution is not very high, so local impacts are more difficult to predict.

Of course CO_2 doubling is an arbitrary point, as in fact it appears likely that this will be reached sometime before the middle of this century, and atmospheric concentration will continue to rise at least until 2100 even with quite stringent measures to reduce emissions. Climate change may not be linearly related to radiative forcing, as it is possible that a large enough perturbation could cause changes in global circulation with unforeseen major impacts. It has been forecast that in the very long term a maximum increase in temperature of 10–18°C would be expected, by about the year 2300, making most of the Earth uninhabitable by today's terms (Cline 1992). This could be a long way off, but will occur unless fossil fuels stay in the ground and are never used.

There are a number of different global models used to predict climate change, developed by research centres around the world, which do not always agree on all aspects. Regional variation, changes in precipitation and seasonal variations tend to vary between models due to their different assumptions. Models are constantly improving as more data are amassed on areas such as ocean circulation and temperature, and our understanding of the Earth's climate improves. The current generation of models includes effects such as the cooling due to SO_2 aerosols, and many of the feedback effects discussed on pp. 207–10, including cloud physics and biofeedbacks such as the effects of changing vegetation cover and phytoplankton.

Table 5.3 summarises the results of the main scenarios used by the Intergovernmental Panel on Climate Change (IPCC) to forecast climate change (IPCC 1995). These are changes by 2100, and as such are dependent upon both the emissions forecast used, and the climate model to predict their impact. The term climate sensitivity is here used to mean the impact upon temperature of a doubling in CO_2 levels.

Other features of climate change that are agreed with some confidence include that warming will be greatest in mid-latitudes, with little summer warming at the poles, and winter land temperatures will rise more than sea temperatures. The strengthened hydrological cycle will lead to wetter winters in high latitudes, but could mean either increases or decreases in both droughts and floods in different regions. Increased incidence of storm force winds is expected in many regions. Most desert regions are projected to become hotter but not wetter and so are likely to grow. Other effects include changes in snowmelt and glaciation which can be

Table 5.3 *Summary of IPCC model predictions of climate change for 2100*

Scenario	Description	Temperature increase, °C	Sea level rise, cm
Low	Emissions stabilised at 1990 levels, low economic growth, low climate sensitivity, strong mitigation from SO_2 aerosols	1	15
Best estimate	Moderate emission controls and best guess assumptions	2	50
High	Emissions continue to grow, high economic growth, high climate sensitivity	3.5	95

modelled separately and whose detailed consequences can be important. For instance Rango (1995) predicts that earlier snowmelt in the USA will reduce freshwater availability in the summer when water demand is highest.

Much of the focus of the modelling effort is to increase understanding of regional variations, and changes in precipitation. For instance, in the UK it is expected that winters will be wetter while summers will be drier in the south and wetter in the north; however some models predict wetter summers throughout the UK, and the magnitude of these changes is not well agreed. It is evident that as strong regional differences occur on a spatial scale of less than a few hundred kilometres, very high resolution models will be needed, which will therefore be data-hungry and demanding of huge processing power. One approach is to model different areas with different resolution, for instance the oceans and deserts could be in much larger grid squares than populated regions.

Validation of models of climate change

Comparing the climate record of the Earth with that predicted by the models allows us to see whether the models work properly (model validation), and also whether the predicted warming is actually measurable now.

For recent times, temperature from historical records at various points around the world can be compared with measurements of CO_2 concentration. Sea surface temperatures are now monitored by remote sensing to produce temperature maps (Plate 5.3) which can be compared for subsequent years to detect changes. Reliable land-based temperature records go back about 150 years, whilst CO_2 has been measured regularly and directly for about 40 years at the observatory at Mauna Loa, Hawaii. Temperatures and concentrations before then can be found from other evidence with reasonable accuracy (Mannion 1999).

Over a longer timescale, records from Arctic ice-cores can be used to measure greenhouse gas concentrations and mean temperature for the last 160,000 years (Figure 5.12). It is found that CO_2 and methane concentrations in air trapped in the

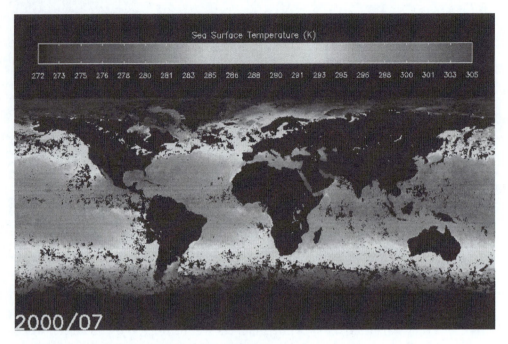

Plate 5.3 *Sea surface temperatures for July 2000.*
Source: reproduced by courtesy of Rutherford Appleton Laboratory

cores correlate closely with temperatures, as expected. Note that present day CO_2 level is far higher than anything over this historical period.

It looks as though temperature is linked to CO_2 concentrations, and that both have increased recently, but how do we know it isn't some statistical fluke? There are some experts who believe that the observed warming is either a statistical anomaly, or can be explained by natural processes. A historical correlation between CO_2 and temperature does exist, before any anthropogenic interference, but this does not prove causality. It could be that lower temperatures caused low CO_2 levels rather than the other way around, as more CO_2 was dissolved into the oceans, or some other biological feedback occurred. The more recent increases in temperature, over the last twenty years, are well within the range of natural variation over the course of centuries and could be just a natural warm phase. Likewise, the apparent increase in storms and other extreme events could be a fluke, or due to biased statistics as more observations are being made now than in the past.

Correlation can be used to show whether the relationship between CO_2 and mean temperature is statistically significant. However this is not totally straightforward, as many other factors influence mean temperature. The most important are:

● solar activity: temperature increases when sunspots increase solar radiation;
● volcanoes: dust in the atmosphere from Mount St Helens and Mount Pinatubo reduced radiation reaching the Earth's surface and had a cooling effect;

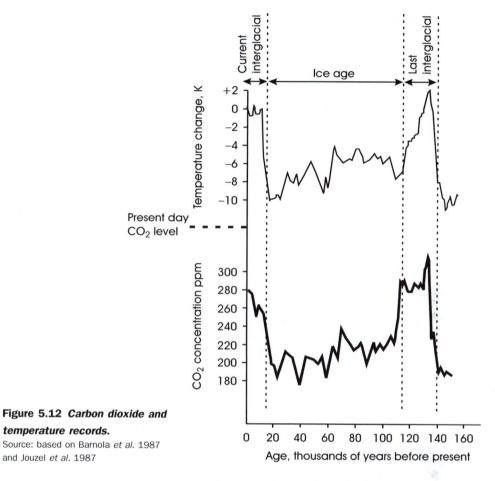

Figure 5.12 *Carbon dioxide and temperature records.*
Source: based on Barnola *et al.* 1987 and Jouzel *et al.* 1987

- aerosol pollution and smoke: the cooling effect of SO_2 and particulates in the atmosphere;
- the Second World War produced enough smoke and pollution to reduce temperatures noticeably;
- urbanisation: many weather stations near urban areas have become warmer because of the growth of cities;
- El Niño: years when El Niño exists are warmer globally; and
- long-term trend: there may be some underlying, long-term trend in temperature due to unknown natural climatic variation.

Using statistical analysis, a correlation can be found between temperatures and these factors, together with CO_2 concentration, that is statistically significant and corresponds with the known physical mechanisms. This gives an equation that predicts temperature as a function of these variables. The expected temperature can then be calculated, according to the observed CO_2 increase, and compared with actual measured temperatures. This process is not straightforward, as the aim is to combine all available knowledge about climate change mechanisms with statistical analysis for random variations. This is being done on a regional basis, with vertical and seasonal temperature variations also being analysed. By this method it can be shown that there is a good agreement between actual and modelled temperatures taking warming from observed increases in CO_2 into account. In particular, in the years since 1980 there has been continued warming that is not explained by any other cause, but that fits well with the increase due to the greenhouse effect. While statistical analysis can never provide 100 per cent certainty, the vast majority of those working in the field now accept that some of the observed temperature change is due to CO_2. To quote the IPCC (1995), 'the balance of evidence suggests that there is a discernible human influence on climate', a statement which took hundreds of hours of international negotiations to agree. In fact it is estimated that global mean temperatures have now risen by 0.3–0.6°C since pre-industrial times, as a result of the greenhouse effect. In other words, global warming has started and can be observed.

Summary

- The general atmospheric circulation consists of cells with winds driven by convection and momentum, in a fairly regular pattern. It serves to redistribute heat energy, and provides the conditions for secondary weather patterns such as low-pressure cyclones to develop.

- Within the oceans, currents are also driven by solar input, determined by temperature changes, thermohaline currents and surface winds. These are important in determining ocean temperatures at depth and in mixing dissolved gases, as well as affecting climate.

- El Niño, a periodic reversal of the normal current system in the South Pacific, is thought to be appearing more frequently, affecting the climate of countries throughout the southern hemisphere.

- Formation of holes in the ozone layer at the poles has allowed more damaging UV radiation to penetrate the upper atmosphere. Ozone has a small direct influence on climate, but there are feedback effects such that global warming may worsen the ozone holes.

- The greenhouse effect is caused by changes to the Earth's radiative balance as a result of increases in atmospheric concentrations of CO_2 and other greenhouse gases.

- Climate change, thought to be as a result of these emissions, has been observed in the form of increases in global mean temperature and increasing sea levels, and is expected to include increased frequency of storms and extreme events.

- Future changes in climate are subject to uncertainty, due to the many feedback effects involving precipitation, albedo, vegetative cover, cloud cover and sea surface temperatures.

- Climate models have been validated with sufficient confidence for the majority to believe that, without stringent cuts in emissions, further climate change will occur and will be highly damaging in terms of both ecosystems and economics.

Questions

1 What effect would desertification have on albedo of an area of land? What effect would this have on temperature?

2 What common atmospheric gas absorbs light with a wavelength of 1.4 μm?

3 What would be the effect on the absorption spectrum of the atmosphere, of water vapour condensing to form clouds?

4 Why is the greenhouse warming potential for nitrous oxide similar whether expressed over 20 or 100 years?

5 What is the increase in radiative forcing brought about by doubling CO_2 concentrations from 280 ppm to 560 ppm? What would be the change due to the same increase again, from 560 ppm to 840 ppm? Explain.

6 What short-term effect will controls on acid precipitation, reducing emissions of sulphur dioxide, have on climate change? Will this effect be relevant in the long term?

Answers to numerical parts

5 4.37 W m^{-2}; 2.55 W m^{-2}

Further reading

Atmospheric Processes and Systems. R. Thompson. Routledge, London, 1998. Covers all the main aspects of climate and weather systems in a concise, readable style.

Natural Environmental Change. A. Mannion. Routledge, London, 1999. Includes details of climate records and the various methods used to uncover them.

General Climatology (4th edn). H. J. Critchfield. Prentice-Hall, New Jersey, 1983. A comprehensible text covering climate and weather.

An Introduction to Atmospheric Physics. R. G. Fleagle and J. A. Businger. International Geophysics Series, Vol 25. Academic Press, London, 1980. Quite detailed coverage of climate processes and radiation in the atmosphere.

Physics of the Environment and Climate. G. Guyot. John Wiley & Sons/Praxis Publishing Ltd, Chichester, 1998. Covers a wide range of physical and climate processes, with an agronomical perspective.

IPCC second assessment report: Climate Change. The Science of Climate Change. J. T. Houghton, L. G. Meira Filho, B. A. Callender, *et al*. (eds). Contribution of Working Group 1 to the 2nd assessment of the IPCC. Cambridge University Press, Cambridge, 1995 (also available online at http://www.ipcc.ch), and: *Climate Change: The IPCC Scientific Assessment*. J. T. Houghton, G. J. Jenkins and J. J. Ephraums. Cambridge University Press, Cambridge, 1993. These provide a consensus view of scientific issues surrounding climate change.

Oceanography – An illustrated guide. C. P. Summerhayes and S. A. Thorpe. Manson Publishing, London, 1995. Useful information on ocean currents.

6 Sound and noise

We are all subjected to increasing levels of environmental noise, from traffic and aircraft, from industrial processes, in the workplace and due to urbanisation. Some noise is merely annoying, while in other cases it can be hazardous to hearing, cause stress or prevent us from enjoying our normal daily lives. For this reason noise pollution is now recognised as a serious and increasing environmental issue. The following key concepts are covered in this chapter:

- **Sound is a pressure wave consisting of vibrations within a wide audible frequency range, whose characteristics depend on wave properties**
- **The decibel scale is a logarithmic scale used to measure sound**
- **Noise can be defined as unwanted sound, with noise pollution being a serious environmental and health problem**
- **Human perception of sound is a non-linear response that varies with the individual, while the impact of noise pollution is even more variable**
- **Policies and practices in noise control reflect both the physical characteristics of sound and the human responses**

Measuring noise is not entirely straightforward, because of the various characteristics of sounds of different frequencies, the logarithmic decibel scale used for measurement, and the human response to sound. Sound has many properties such as frequency and wavelength, absorption, reflection, refraction that are typical of wave motion, an understanding of which aids in both monitoring and controlling noise.

Sound waves

Sound waves are pressure waves in the air caused by the vibration of a solid object. For instance, when a guitar string is plucked, it vibrates at a certain frequency. This moves the air around it, and a sound wave is emitted.

Sound waves are compressional, longitudinal waves – they consist of variations in pressure moving in the same direction as the vibration. By contrast, waves on the sea and electromagnetic waves are transverse waves where the wave moves at right angles to the direction of the oscillation.

Like any other wave, sound waves have a frequency and a wavelength. These will depend on the properties of whatever produced the sound, e.g. a longer piano string generally produces a longer wavelength and a lower frequency – a deeper note.

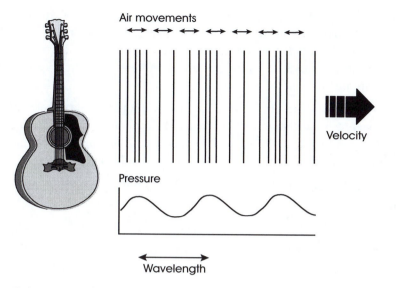

Figure 6.1 *Production of a sound wave.*

Increasing the tension in the string produces a higher frequency of vibration, and so a higher note.

The velocity of sound

All sound waves move at the same velocity in air, about 331 m s^{-1}. This compares with the speed of light of 300,000,000 m s^{-1}, nearly a million times faster.

Velocity is related to frequency and wavelength as for any wave:

$$v = f\lambda \qquad (6.1)$$

As the velocity of sound in air is always the same, this means the longer the wavelength, the shorter the frequency and vice versa. Velocity depends on the medium through which the wave is propagating, its temperature, density, and pressure. In a solid, it also depends on Young's Modulus as wave propagation is related to elasticity. In water, sound travels faster than in air – 1,435 m s^{-1}. Because sound waves depend upon vibrations of a medium (solid, liquid or gas), they cannot exist in a vacuum, or move through one. In Space, as they say, no one can hear you scream.

Music

The nature of sound can be readily understood by looking at music. A musical note is a sound wave of a certain frequency, the musical pitch determined by that frequency. A tuning fork tuned to middle 'A' produces a note of frequency 440 Hz, where 1 Hz is one vibration per second. The audible range is from about 15 Hz to 20,000 Hz. Above 20,000 Hz, sound waves are termed ultrasonic (see p. 224). Below 15 Hz they cannot be heard but may be felt as vibration, if strong enough, like that from traffic.

An octave corresponds to doubling or halving of frequency. So a frequency of 220 Hz is an 'A' an octave lower than the 440 Hz middle 'A' , and 880 Hz is an 'A' an octave higher. Halving the length of a guitar string by pressing a fret doubles the frequency, and so produces a note an octave higher.

The wavelength of a note is given from the formula for the speed of sound in air (Equation (6.1)):

$$\lambda = v/f \quad \text{where } v = 331 \text{ m s}^{-1}$$

So a middle 'A' will have wavelength (in air) of

$$\lambda = 331 \text{ m s}^{-1}/440 \text{ Hz}$$
$$= 0.75 \text{ m}$$

This is the distance between successive vibrations as the sound wave comes towards you. A high 'A' would therefore have a wavelength of half this (0.375 m) and a low 'A' would be twice this (1.50 m). The amplitude of the wave determines how loud the note is – if you pluck a guitar string gently, it will vibrate with a smaller amplitude, and produce a quieter note. The pressure variations in the wave formed will be less, but the frequency and wavelength are unaffected, and so the pitch is the same. These characteristics are illustrated in Figure 6.2.

As well as these simple characteristics, the sound of any musical instrument will have a distinctive waveform from secondary waves. Any real instrument produces a range of different frequencies all at once, dominated by the frequency of the note being produced. So the result is a waveform with a general shape of the dominant frequency, but superimposed with wobbles and spikes according to the instrument. A very pure sound like a tuning fork will have a neat, smooth waveform. A coarse sound like a badly tuned radio or a guitar with a fuzzbox produces a very irregular waveform, and human speech even more so with frequency changing all the time. However even non-musical sounds usually have a characteristic frequency, which is

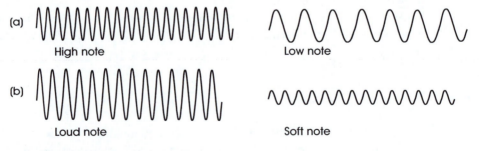

Figure 6.2 *Characteristics of simple sound waves.*

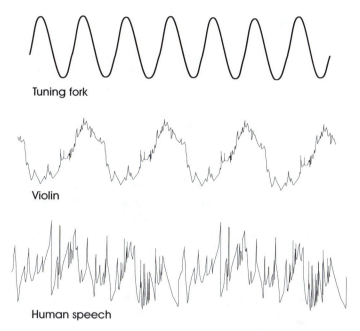

Tuning fork

Violin

Figure 6.3 *Waveforms of various sounds.*

Human speech

the frequency that is most dominant, around 1,000–2,000 Hz for speech. True white noise comprises random noise over a wide frequency range with no characteristic frequency.

When you listen to music, your ear and brain separate the different frequencies to identify the characteristics of, say, a violin. Not only that, but when listening to a band, you can separate out all the different instruments, and understand the words and tell who the singer is too – all from the vibrations of your eardrum.

Mathematically, a similar job can be done on any waveform by means of the Fourier transform. This is a mathematical technique that expresses any waveform as the sum of components contributed at many different frequencies. It provides a general tool for analysis and characterisation of any sound, or any other wave for that matter. The sound can be represented as a spectrum showing the level at each frequency, in a graph of intensity against frequency as shown in Figure 6.4. This type of analysis can be relevant to scales of measurement and to noise control.

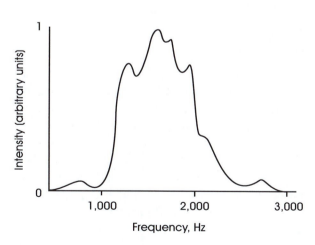

Figure 6.4 *Frequency spectrum for typical human speech.*

Ultrasound

Ultrasound is inaudible sound at frequencies above the range of human hearing, that is above about 20,000 Hz, as used in ultrasound scanners. Many creatures can hear sounds that are too high-pitched to be audible to humans, including bats and small birds. Birdsong sounds even more intricate to the bird, as it can hear many high frequency notes that we miss.

Bats use ultrasound for navigation in the dark, very much like radar or a ship's echosounder. They emit continuous high pitched squeaks, and by listening to the echoes can detect the presence of objects that reflect sound, gauging the distance according to the time taken for the echo to return. By this means bats can fly with remarkable agility around trees or buildings, and locate their insect prey in pitch darkness.

Ultrasound is suitable for this purpose because it has good resolution, due to its short wavelengths (1–2 cm), which reduces the amount of diffraction, and it reflects well. Lower frequency sound would not work, as it would travel through many obstacles without being reflected, and when echoes were produced the longer wavelengths would make accurate measurements impossible. Radar by analogy uses electromagnetic waves with wavelengths of a few centimetres in a very similar process.

Propagation of sound and acoustics

Acoustics is the study of how sound travels around in buildings and other spaces. When an orchestra produces a note, it will travel around the concert hall, being reflected from some surfaces and absorbed into others. What you hear is a combination of sound reaching you directly and from reflections or echoes. Some surfaces, particularly wood and soft furnishings, absorb more of the high notes, so the reflected sound will be lower in pitch and soften the overall sound. Hard concrete and stone reflect precise echoes and don't absorb much.

Different materials will transmit or absorb sound waves of different frequencies. In general, deeper notes with low frequency and long wavelength will be transmitted more effectively, while higher notes are absorbed more, for instance by soft furnishings. So if someone is playing loud music in the house next door, all you may hear is the bass line and the drum beat, with the higher notes absorbed by the wall. This is because the deep notes have wavelengths that are long compared with the width of the wall – up to 8.3 m for a 40 Hz low 'E' from a bass guitar. So they go straight through. By contrast, the wavelengths of tens of centimetres for the higher notes are similar to the width of the wall and so will be absorbed.

If a large concert hall has hard, unbroken, surfaces, the result can be an echoey sound – like in an empty church. Absorption of sound will reduce the overall

Box 6.1

Natural acoustics

Different natural environments also differ in their acoustics, and some species have evolved to adapt to specific conditions. Rainforests consist of many tree and plant species with hard, reflective, waxy leaves, which makes for a lot of echoing background noise. To cope with this, typical bird and animal calls are quite deep and musical, on single sharp notes, that carry well. In an English deciduous woodland by contrast, the softer leaves and more open canopy have quite different acoustical properties, and melodious, complex birdsong at high frequencies will carry well without being lost. In a desert, high pitched scratchy croaks and chirps will carry as there is less to absorb them. Other species have calls designed to carry well over very long distances, such as the bittern's 'boom' or frogs' throaty resonant croaks, using single, deep notes that are less easily absorbed by their wetland habitats. Alarm cries, like a bird's high-pitched, short cheep or a rabbit stamping its rear feet, are designed to alert others without being easily located, putting the alarm-giver at risk. Underwater, certain sounds can carry very long distances. Species such as whales and dolphins can communicate at distances up to hundreds of kilometres through their 'songs', typically low frequency, long smooth sounds that carry best.

volume as well, so large spaces with hard surfaces can be very noisy. A smaller room or one with a lot of soft furnishings absorbs more and sounds more rounded. For instance, as people and their clothes absorb sound, particularly higher notes, your hi-fi will sound different when you test it before a party than later on when everyone's got there. In a room full of people, the higher frequency sounds from the music are absorbed, resulting in lower volume and more bass. To compensate, you'll have to increase the treble and turn the volume up.

A *New Scientist* survey (*New Scientist* 2000) of restaurant noise found noise levels significantly higher in large, modern, 'concrete and chrome' restaurants than in the old fashioned variety featuring carpet and flock wallpaper. Noise levels up to 97 dB in the former came about, a level comparable with the loudest industrial environments, where ear protection would be required for workers. These levels were without music playing, as once the ambient noise is high enough to make it difficult to hear conversation (at around 80 dB, see pp. 233–4), people talk more loudly until everyone is shouting. Soft furnishings absorb enough sound to keep conversations private and audible to the participants.

The acoustics of a room also depend on resonant frequencies of objects in it. If the resonant frequency of an object coincides with that of a sound, it will make it vibrate, amplifying the sound. For instance if a tuning fork is held against a table, the note produced sounds louder as the table vibrates. Resonant frequencies depend on the size, shape and hardness of objects. If some feature has a strong resonance at one frequency, it can distort the sound by amplifying or by absorbing that frequency.

The Doppler effect

When a fire-engine drives past, the sound it makes appears to change as goes by, becoming deeper in pitch:

NEE-NAA-NEE-NAA-NEE-NAA-NEE-NAO-NOO-NAW-NOO-NAW-NOO-NAW
      coming towards you   passing   going away
     wavelength compressed   wavelength stretched
        frequency higher   frequency lower

This is because of the relative motion of the vehicle, changing from coming towards you to going away. When the vehicle is moving towards you, the wavelength is slightly shortened on each vibration, by the vehicle having moved a short way, making the frequency higher. When it is travelling away, the wavelength is lengthened and the frequency becomes lower. Because the speed of the vehicle is significant when compared to the speed of sound, the effect is noticeable. This is known as the Doppler effect.

Light also behaves in the same way, but this is not commonly observed because the object has to be moving at a speed that is a significant fraction of the speed of light – fire-engines do not go that fast. This Doppler shift in light frequencies is important in astronomy, to measure the motion of stars, and in other fields.

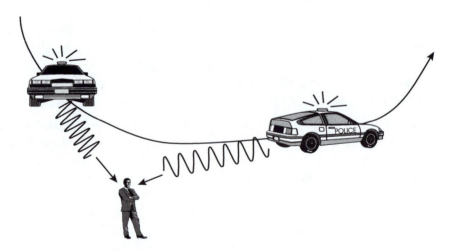

Figure 6.5 *The Doppler effect.*

Measuring sound: the decibel

Sound is measured in decibels, named after Alexander Graham Bell. One bel is defined as the log of the ratio of the intensity of two sounds:

$$B = \log(I/I_0) \tag{6.2}$$

Here I is the intensity of the sound, defined as the energy carried per square metre of wavefront. I_0 is a baseline intensity to which the sound is compared, generally 10^{-12} W m^{-2}. A decibel is a tenth of a bel, which gives a more convenient unit. So a decibel is defined as:

$$L = 10 \log (I/I_0) \tag{6.3}$$

The decibel scale is logarithmic – increasing the decibel level by 10 results in a sound 10 times as loud, increasing it by 20 results in a sound 100 times as loud and so on. Doubling of intensity corresponds to an increase of 3 decibels. A logarithmic scale is used for two reasons. First the range of sound intensities that is audible is very wide, from intensities of 10^{-12} W m^{-2} to 10^{-5} W m^{-2}, which in logarithmic terms is a much narrower range, from 0 to 140 dB. Second, to a broad approximation, our perception of loudness of sound resembles a logarithmic scale, particularly over a central frequency range.

A louder noise will carry more energy, and the variations in pressure in the sound wave which carry this energy will be greater. The decibel can be expressed in terms of the sound intensity (as shown above), or as the energy in the sound wave, used in calculations of noise at source, or the sound pressure level, which is generally the quantity measured in a sound meter. All are measured relative to a given baseline – the scale is not absolute, but defined as the difference between two sounds. The baseline is the quietest sound audible, which has intensity of 10^{-12} W m^{-2} at 1,000 Hz, or energy of 10^{-12} W, or pressure of 20 μPa. In the case of sound pressure levels, a decibel is defined as $20 \log (P/P_0)$ where P_0 is 20 μPa. The figure 20 replaces the 10 in Equation (6.3), because intensity (or energy) of the wave is proportional to the square of pressure. Thus the three definitions of decibels are equivalent:

$$L = 10 \log (I/10^{-12}) \quad \text{in terms of intensity in W m}^{-2}$$
$$= 10 \log (W/10^{-12}) \quad \text{in terms of energy in W}$$
$$= 20 \log (P/20 \times 10^{-6}) \quad \text{in terms of pressure in Pa}$$

Combining decibel levels

If a number of different sounds are all being made at once, how do you find the total sound level? Because of the scale being logarithmic, you cannot just add up decibels like apples and oranges. What you would need to do is convert them back to pressure levels by taking powers of ten. Rather than this, you can do it graphically, using the following method.

1 Start with the two lowest levels.
2 On Figure 6.6, find the difference between the two levels on the horizontal axis.

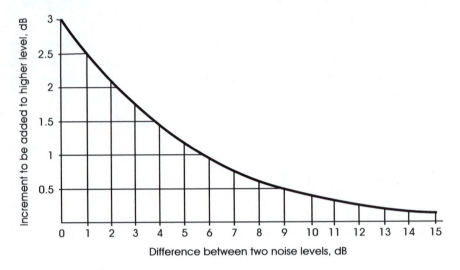

Figure 6.6 Decibel addition chart.

3 Find the resulting increment in decibels on the vertical axis.
4 Add this to the higher of the two levels to give the combined level.
5 Repeat steps 2–4 for this combined level, together with the next loudest sound.
6 Round off to the nearest whole number.

For example, in a noisy street you are subjected to the noise from two cars at 75 dB each, and a lorry at 80 dB. First add the two cars – as the difference is zero, add 3 dB from the graph, to make 78 dB. Then add the lorry – the difference is 2 dB, so from the graph the increment is 2.1 dB, making 82.1 dB.

Mathematically, the same result can be obtained by converting back to sound intensity levels, where total sound intensity is I_{tot}, relative to the baseline I_0, and total level in decibels is L_{tot}:

$$I_{tot}/I_0 = 2 \times 10^{75/10} + 10^{80/10}$$
$$= 163 \times 10^6$$
$$L_{tot} = 10 \log (163 \times 10^6)$$
$$= 82.1 \text{ dB}$$

This process is only valid for sounds that are independently generated and therefore the frequencies randomly interfere either constructively or destructively. For the case of two sounds that are strongly correlated, for instance two radios tuned to the same station, the sum will be more than this as interference between the waveforms will be constructive. The converse of this situation, destructive interference, can be used in noise reduction (pp. 236–7).

It is not valid to take a simple average of sound decibel levels. For instance, an environmental consultant may wish to monitor the noise from an industrial plant.

They would take several measurements of noise level L at different times, and at different locations inside and outside the plant. To find a time-weighted average for each location, they would need to use the formula:

$$L_{eq} = 10 \log \frac{\sum 10^{L/10}}{N} \tag{6.4}$$

This average, L_{eq}, represents the equivalent continuous sound level over a stated period. To avoid the rather confusing averaging process, noise levels from the sample data are often expressed as the percentage of time a certain level is exceeded, rather than the average. For instance the noise level that is exceeded in 10 per cent of sampled readings is termed L_{10}, while the level exceeded in 90 per cent of the samples is known as L_{90}, representing typical values for peak and background noise respectively.

Propagation of sound over distance

When sound is produced from a point source (such as a building site) it spreads out in three dimensions – outwards and upwards. The intensity of the sound will decrease further away from the source, according to an inverse-square law. The intensity is inversely proportional to the square of the distance from the point.

$$I = W/(4\pi R^2) \tag{6.5}$$

where W is the sound energy of the source, R is distance away. The $4\pi R^2$ represents the area of a sphere, over which the noise is spreading.

To convert to decibels, divide by I_0 and take logs, then multiply by 10 and rearrange:

$$10 \log (I/I_0) = 10 \log [W/(I_0 4\pi R^2)]$$

$$L = L_w - 10 \log R^2 - 10 \log (4\pi)$$

$$L = L_w - 20 \log (R) - 11 \tag{6.6}$$

In Equation (6.6) L_w is the sound power level at source. This is not usually measured directly, but can be calculated from a noise level reading at a known distance. From the above equation it can be seen that doubling the distance away from a point source will reduce the decibel level by 6 dB, that is reducing it to a quarter of its intensity, because it is based on an inverse-square law.

For a source that is a line rather than a point, such as a road or railway, the sound cannot spread out in this way. In this case the noise can only dissipate upwards and away in two dimensions, so it is spread over a half cylinder shape, above and along the road. So the noise at a distance will be inversely proportional to the distance, rather than its square. This gives the intensity as:

$$I = W/2\pi R \tag{6.7}$$

This can be converted to decibels in the same way as Equation (6.6), to give:

$$L = L_w - 10 \log (R) - 8 \tag{6.8}$$

In this case doubling the distance will decrease the noise level by 3 dB, halving its intensity, as the relationship is linear.

It is evident from the above that when sound is measured, it must be done at an appropriate distance. The noise from a road or any other noise source must be monitored at a stated distance, and the national or international standards for noise pollution must define the distance at which the monitoring is carried out.

These equations assume that noise is spreading equally in all directions, which may not always be the case. Any other obstructions such as buildings or topography can reflect or absorb sound, altering the way it disperses. Attenuation of sound will occur over distance as it is absorbed in the atmosphere, an effect which is greater at high frequencies. Corrections can be made for this attenuation by use of tables. Atmospheric effects can also affect sound transmission. You may have lain in bed at night listening to a far-off road, railway, church bell or fog horn that is quite inaudible during the day. You can hear them at night partly because of the lower background noise, but also because of atmospheric effects. At night, the atmosphere cools close to the ground, which commonly produces a layer of warm air overlying the cold close to the ground. Sound waves are refracted by this layer because of the difference in density, resulting in an increased velocity in the colder air. At a distinct boundary between cold and warm air, sound waves striking at an oblique angle can also be reflected, almost as if off a hard surface. This combination of refraction and reflection causes them to curve down again towards the ground

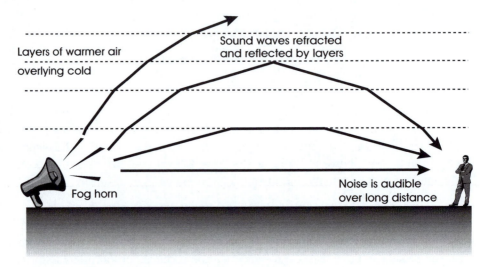

Figure 6.7 *Refraction of sound in the atmosphere.*

(Figure 6.7). The result is that noise can be heard at a much greater distance than during the day, when the temperature profile is the opposite causing sound waves to be refracted up and away.

Noise and noise nuisance

Noise nuisance does not depend purely on how loud a noise is. Some sounds are far more annoying than others – a badly played violin or a baby crying might be perceived as worse than a rumble of traffic of the same loudness. However, it is a fundamental law that everybody's neighbours have bad taste in music!

Regulations governing noise must be based upon objective criteria, hence measurement of the level in decibels is important. Perceived loudness depends upon the individual but most people share a similar auditory response. Hearing damage can come about through environmental or workplace exposure to noise, requiring regulation on noise levels and protection.

Human perception of sound and noise

The human ear is an incredibly sensitive instrument, able to perceive sounds over a huge range of frequencies and intensities, and many animals have hearing that is far more sensitive. In the ear, incoming sound waves are directed through the outer ear to the eardrum, which vibrates. This motion is amplified by three tiny bones in the middle ear, through a membrane to the cochlea in the fluid-filled inner ear. The cochlea is a spiral-shaped organ, that is surrounded by tiny hairs. These hair cells produce signals in the auditory nerves at specific frequencies, which are relayed to the brain. It is these hair cells that can sustain damage when exposed to excessively loud noise, producing partial hearing loss either temporarily or permanently. Very sudden loud noise or explosions can also damage the eardrum.

Loudness level as perceived by any individual will depend upon the frequency of the sound, and is not linearly related to the level in decibels. The peak of loudness is at 4,000 Hz, a frequency that corresponds to a note near the top of a piano's range. For a given decibel level, a sound of this pitch will sound loudest. Loudness level is measured by the phon, which is defined by an international standard found by measuring hearing of a large number of normal, young individuals, illustrated in Figure 6.8. The loudness level in phons is equal to the loudness in decibels at 1,000 Hz (a frequency typical of human speech). The response is fairly flat in the mid-range of frequencies, but for low and high frequencies loudness drops away. A 10 phon response is produced by definition by a 10 dB sound at 1,000 Hz, but at 60 Hz it would take 40 dB to sound equally loud. At louder intensities the response is flatter, with less variation with frequency. This auditory response varies between individuals, and thresholds commonly shift with age or exposure. The young can

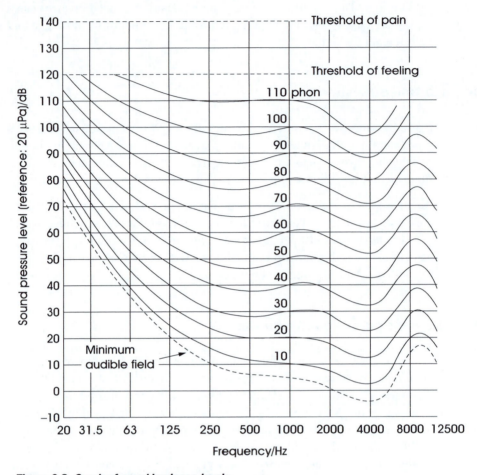

Figure 6.8 *Graph of equal-loudness levels.*
Source: reproduced by permission from ISO (1987) Annexe A p. 4

generally perceive quieter and higher-pitched sounds than the old, while occupational exposure to loud noise of a particular frequency can result in impaired hearing at that frequency.

Apart from physical damage and hearing loss, noise can produce varied psychological responses. Noisy environments have been linked to aggression, for instance a German study by Niebel *et al.* (1993) linked school violence to classroom noise level. If this is the case, it is even possible that better acoustics and more soft furnishings in schools or prisons could reduce violence. Noise can produce physical symptoms such as increased heart-rate or dilated blood vessels, part of the physiological response that evolved when our ancestors needed to be alert to the threat of a hungry beast or other danger. In the modern environment this inappropriate response to unavoidable stimuli just adds to stress levels, with associated health effects.

Noise levels

Noise nuisance falls into a number of categories, in increasing order of severity:

- acoustic privacy;
- sleep interference;
- speech interference;
- hazard to hearing from long-term exposure;
- hazard to hearing from short-term exposure; and
- acoustic trauma – causes pain.

Some noise nuisance may be encountered at relatively low levels of noise. In particular, lost sleep may be caused from fairly quiet but irritating or intermittent sounds – just as you're dropping off, that dog starts barking again or another plane goes overhead. Also if general background noise is low, any additional noise may be considered worse. The noise of wind turbines is very low compared with, say, a busy road, but if you have just retired to enjoy the peace and quiet of the Welsh countryside, you may find having a wind farm humming away at the end of your garden disturbing. Noise nuisance is subjective, as some people are more sensitive or find particular things worse than others.

In a typical office, background noise will be from 50–60 dB. Higher levels of noise may make understanding speech difficult – from around 80 dB. Occupational exposure is limited in Britain to 85 dB for an 8-hour working day – above this level, ear protection or noise reduction should be used. By 95 dB, typical of an unsilenced pneumatic drill at close quarters, there would be a hazard to hearing, and by 120 dB damage to your ears would cause pain. Many common environmental noise problems are between 60–90 dB, such as that caused by busy roads.

While human hearing can detect noise down to about 15 Hz, even lower frequency noise can cause nuisance. Low level vibrations such as those from heavy lorries, trains or mining operations may be inaudible yet still be felt by people, often in their stomachs rather than their ears. Such low frequency noise may cause nausea or headaches as well as sleep loss, and can easily penetrate through walls and other barriers, making it more difficult to control.

Noise measurements

A sound level meter measures the sound level via a microphone and amplifier, which produces an electrical signal, displayed in decibels on the readout.

For the measurement of noise, the decibel scale is adapted to more accurately represent the human perception of loudness. To do this, a sound level meter contains an electronic filter that selectively attenuates certain frequencies. There are three sets of filters commonly used, known as A, B and C, producing the dBA,

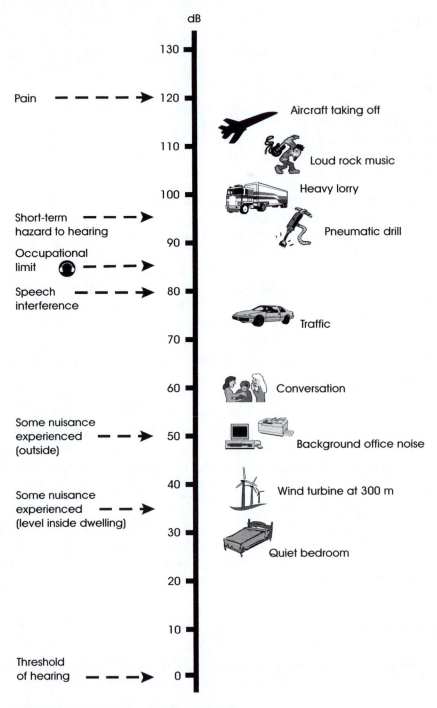

dB

130

Pain ‑ ‑ ‑ ‑ ➤ 120

Aircraft taking off

110

Loud rock music

100

Heavy lorry

Short-term ‑ ‑ ‑ ➤ 95
hazard to hearing

Pneumatic drill

90

Occupational
limit ‑ ‑ ‑ ➤ 85

Speech ‑ ‑ ‑ ➤ 80
interference

70

Traffic

60

Conversation

Some nuisance
experienced ‑ ‑ ‑ ➤ 50
(outside)

Background office noise

40

Wind turbine at 300 m

Some nuisance
experienced ‑ ‑ ‑ ➤
(level inside dwelling)

30

Quiet bedroom

20

10

Threshold
of hearing ‑ ‑ ‑ ➤ 0

Figure 6.9 *Noise levels of common sounds.*

dBB and dBC scales. The A scale filters low and high frequencies to fit a curve that represents the human auditory response at 40 phons. The B curve filters low frequencies less, corresponding to the response at 70 phons, while the C scale is almost flat, more representative of the response for very loud sounds. Of these the dBA is the most widely used, almost universally in noise regulation, despite the fact that the B scale would be more appropriate for the values commonly being measured as noise nuisance.

Impulse sounds, like hammering, may be too sudden to register on a conventional sound meter, as any meter has a response time over which noise is averaged. An impulse meter has a shorter response time, more accurate in measuring sounds of short duration.

Because of the individual nature of the response to noise, no objective measurement can accurately quantify the nuisance value of noise. There are a variety of methods that take into account the length of time a noise level is exceeded, the frequency level or the suddenness of sounds, in an attempt to better represent nuisance value (see Vesilind *et al.* 1988: 482). The psychological response cannot be evaluated in this way – for instance some may object to cockerels and church bells in the countryside, while others enjoy these sounds as part of rural life. The physical characteristics of the sound in this case are virtually irrelevant.

Controlling noise

Methods of controlling noise take advantage of some of its wave properties, including absorption, damping, reflection and interference.

Sound-proofing materials absorb noise over a wide range of frequencies. They are materials like expanded polystyrene, fibreglass or even old egg cartons, that are soft and/or contain air pockets, so absorbing the vibrations. These are used in recording studios to reduce incoming noise, and in noisy developments such as clubs or theatres to prevent excess noise escaping to nearby residential areas. The choice of material will depend upon the frequency range of the noise to be controlled, as certain materials absorb more effectively at certain frequencies. Sound insulation is important in walls and ceilings, particularly between neighbouring properties and in apartment blocks. The sound insulation of a wall, R, can be defined as

$$R = 10 \log (I_i/I_t) \tag{6.9}$$

where I_i and I_t are the incident and transmitted sound intensities (Boeker and van Grondelle 1999). This gives the difference in decibel levels at each side of the barrier. It will be reduced in the case of sounds transmitted directly to the wall rather than through the air, for instance where residents in the flat upstairs have their hi-fi speakers on the floor. Sound insulation is frequency dependent. To be effective, sound-proofed areas must be solid and without cracks, as sound will be transmitted effectively through quite small holes or cracks (such as those under doors) reducing any noise insulation significantly.

In the workplace, sound-proofing takes the form of ear protection (e.g. ear muffs) for the staff involved with noisy equipment. Mufflers can be used on equipment such as pneumatic drills to reduce both environmental noise and damage to the operator. Industrial equipment can be mounted on springs or on elastic rubber mountings, to prevent sound transmission through the floor, particularly effective in damping low frequency vibration. This principle has even been used in buildings, resting the entire building on large springs, to prevent noise and vibration transmitted through the ground from a nearby railway.

Resonant frequencies may be useful – a damping mechanism on a drill for instance works by having a resonant frequency very different from that of the noise the drill makes. The noise from the drill thus finds it hard to make the baffle vibrate and the sound wave is damped or attenuated.

Sound barriers are often used alongside motorways in urban areas. They act by reflecting the sound back into the motorway and upwards, away from surrounding housing. They are simply tall, rigid, solid fencing of wood, metal or plastic. They are of limited effectiveness, as sound at the top of the barrier will be diffracted back downwards into adjacent gardens.

A more radical solution is to reduce noise at source. This means designing cars, lorries, planes, industrial machinery and even garden equipment to make less noise,

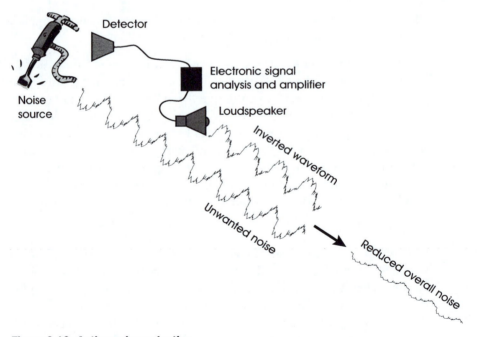

Figure 6.10 *Active noise reduction.*

> ## Box 6.2
>
> ### Owls: nature's methods of noise control
>
> Owls' silent flight gives them a great ability to surprise their prey without audible warning, as well as meaning they can hunt by sound as well as sight – they can hear the rustling of a moving mouse without it being drowned out by their own wingbeats. How do they do this?
>
> Three main components contribute to a bird's flight noise: airflow caused by the downbeat of the wing; turbulence around the wing, and movements of feathers relative to one another. In an owl, large wings mean slow, gentle flapping, reducing the first noise source. The design of the wing feathers includes a stiff leading edge and a much softer trailing fringe, which greatly reduces turbulence around the wing. Finally, the upper surface of the feathers has a soft downy covering that reduces inter-feather noise. The result is their virtually silent flight.

which has been done to some extent. Noise reduction can involve reducing vibrations, controlling combustion carefully, improving aerodynamics, reducing turbulence, avoiding resonant frequencies and more fundamental design changes. Many of these changes will also improve efficiency and add to the equipment's life. Much of vehicle noise comes from the contact with the road surface, but both tyre pattern and road surface material can be chosen to minimise noise. Speed limits and lorry bans also help reduce road noise in urban areas.

A final option may be active noise reduction. This is a technique to reduce noise levels by producing an identical noise that is exactly out of phase with the offending noise waveform. Destructive interference occurs and the resulting noise amplitude can be reduced by as much as 15 dB. The technique involves measuring the noise and sending the signal through an electronic system that analyses the waveform and reproduces it in inverted form, that is exactly out of phase with the first (see Figure 6.10). The signal is then sent to an amplifier and loudspeaker, to rejoin the initial noise and have its 'cancelling out' effect. In order to be effective the signal must be very precise in both waveform and phase, otherwise the overall noise may increase.

Noise contours

Noise contours provide a planning tool of use in designing development such that noise nuisance is avoided. The noise contour like any other contour is a line joining points of equal value, in this case equal predicted (or measured) noise level.

For example, in the environmental impact assessment of new wind farms, it is common to include a map of the site with noise contours. From a knowledge of the

Box 6.3

Airports and aircraft noise

Airport noise is distinguished by being discrete rather than continuous, as individual aircraft take-off and land. An aircraft taking off can produce as much as 110 dB measured at ground level at monitoring stations around airports. Highest volumes are during take-off as this is when most energy is being expended. With increasing air traffic, noise from aircraft is an increasingly important issue, particularly to those living close to existing or proposed airports. The high level of noise commonly causes sleep problems, prevents normal conversation and vibrates buildings in affected properties close to flight paths. Loss of sleep is of specific concern, such that at many airports stricter controls apply during the night.

Aircraft noise is not only very loud, it covers a wide frequency spectrum, being strong from 50 Hz to 5,000 Hz. A specialised decibel scale, the dBD scale, has been developed to approximate the response to aircraft noise (Vesilind *et al.* 1988). In the UK, a similar scale is used, weighted by frequency to modify the decibel level to represent noise annoyance more accurately, termed the **perceived noise level** and measured in PNdB.

Control by barrier methods is not applicable to aircraft noise, apart from sound insulation of homes. Regulatory controls include the direction of flight paths away from the most populous areas, restrictions on numbers of flights and night-time curfews on airports. Noise certification of aircraft involves comparing measurements of noise with maximum limits, at specified points on the take-off or landing path and to each side, at set distances. As aircraft exceeding these limits are subject to stricter restrictions, this has given manufacturers the incentive to produce quieter jet engines and aircraft.

Supersonic aircraft are louder even when flying at sub-sonic speeds (as anyone living under Concorde's flightpath will testify). Once the speed passes 331 m s^{-1} and the sound barrier is broken, a loud sonic boom is produced due to the aircraft flying faster than the pressure waves produced at bow and stern can travel. This is an insurmountable problem, such that supersonic flight is normally only permitted over oceans.

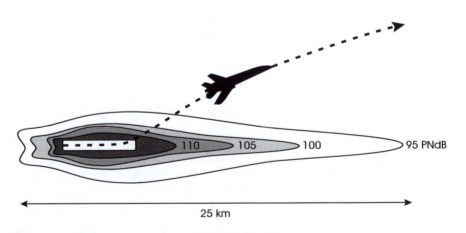

Figure 6.11 *Noise footprint from an aircraft taking off.*

For an individual aircraft, the noise production at take-off and landing can be represented as a noise footprint. This consists of noise contours around the runway depicting maximum noise levels during the take-off. The noise footprints vary for different aircraft, used in prediction of noise levels according to flight schedules.

Noise contours around existing or proposed airports are calculated from data on numbers of flights and flightpaths for various aircraft. Noise production for any aircraft make is known, extrapolated to ground level according to distance and height using physical models of noise dispersion. These contours define areas in which home owners may be liable for compensatory payments to provide noise insulation, or in which new developments of housing would be avoided.

technical characteristics of the wind turbines and wind conditions, the amount of noise at source can be predicted. This can be combined with topographical data and the wind direction to calculate and map noise contours under various sets of conditions. Maps can then be produced, showing how any dwellings would be affected, and ensuring that these predicted noise levels are below regulatory maximums.

Summary

- Sound and noise are longitudinal pressure waves characterised by frequency, and a constant speed in air of 331 m s^{-1}.

- Sound intensity is measured using the decibel scale, a logarithmic scale, useful to cover the very wide audible range, which requires special mathematical methods for adding sound levels or taking averages.

- The distance from the sound source is important in taking measurements as noise dissipates over distance.

- Human perception of sound is a non-linear response, with perceived loudness level measured in phons.

- The dBA, dBB and dBC scales filter certain frequencies to more accurately represent human perception of loudness.

- Noise control makes use of physical characteristics of sound including absorption, reflection, resonance and destructive interference.

Questions

1　Middle C has a frequency of 330 Hz. What would be its wavelength (in air)? What note would have a frequency of 660 Hz?

2　If you take your portable radio on a picnic, you may find it sounds different when you play it in a field to when you play it at home. Why is this?

3 Standing in the street, I am exposed to several sounds simultaneously: people talking at 55 dB, two cars passing at 7 m away making 75 dB each, and a lorry at 80 dB. What is the total noise level I experience?

4 Monitoring noise nuisance levels at sites in residents' gardens close to the A69, I have taken the following readings: 62 dB, 75 dB, 67 dB, 58 dB, 74 dB. What is the average noise level?

5 If a pneumatic drill makes 95 dB of noise when measured at 7 m away, what would be the noise level at 50 m distance?

6 If a busy road has an average noise level of 95 dB when measured at 7 m, what would be the noise level 50 m away?

7 Low frequency road noise is measured at 75 dBA in the garden of a nearby house. If measured on the dBB or dBC scales, would you expect the reading to be higher or lower?

Answers to numerical parts

1   1 m

3   82 dB   4   69.5 dB

5   78 dB   6   86 dB

Further reading

T237 'Environmental control and public health': units 11,12 and 13. Open University, Milton Keynes, 1993. Texts for three units of this course covering noise, noise control and legislation in detail, with an easily understandable presentation.

Introduction to Environmental Engineering. M. L. Davis. McGraw-Hill, New York, 2000. Contains a general introduction to noise and noise control.

Environmental Engineering (2nd edn). P. A. Vesilind, J. J. Peirce and R. F. Weiner. Butterworths, Massachusetts, 1988. Contains a chapter on general and engineering aspects of noise control.

Environmental Physics (2nd edn). E. Boeker and R. van Grondelle. John Wiley, Chichester, 1999. Contains a chapter on acoustics and noise pollution with more mathematical detail.

7 Radioactivity and nuclear physics

Radioactivity is not just about nuclear power stations. Radioactive emissions arise from many activities, from hospitals to coal mining, natural leakage from rocks to household refuse. Apart from these potential hazards, radioactivity has many uses in monitoring and understanding of the environment. The following key concepts are covered in this chapter:

- Nuclear reactions depend upon the binding energy in the atomic nucleus
- Ionising radiation has sufficient energy to ionise target molecules, causing chemical changes and cell damage
- Radioactive isotopes are those which undergo radioactive decay, at a rate measured by their half-lives
- Fissile isotopes may take part in fission reactions such as that used in nuclear power, releasing large amounts of energy
- The design of nuclear power stations and other installations in the nuclear fuel cycle aims to keep both routine emissions and risk of accidental release within safety standards, requiring detailed knowledge of the nuclear reactions occurring
- Fusion reactions involve two light nuclei such as hydrogen combining into one, with possible future potential for power production

Carbon dating and other forms of radiometric dating are vital tools to look at prehistory; radioactivity in plants and algae is used in biomonitoring; radioactive fallout from atmospheric nuclear testing can be used to measure soil erosion; radioactive tracers are widely used in medicine and radioactivity is used in all sorts of devices from X-rays to smoke alarms. In this chapter we will start with some basics of nuclear structure and processes, giving details on some environmental uses and hazards of radioactivity, and move on to cover the principles of nuclear power.

Nuclear physics

Nuclear physics comprises the study of the nucleons (neutrons and protons) that make up the nucleus of the atom, and the forces and energies between them.

The structure of the atom

Any atom consists of a nucleus containing neutrons and protons surrounded by electrons.

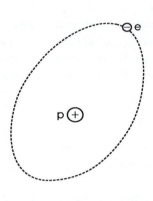
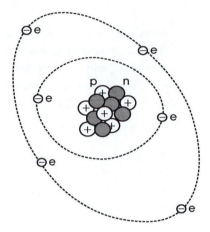

Figure 7.1 *The structure of the atom.*

	Hydrogen atom, $_1^1H$	Carbon-12 atom, $_6^{12}C$
Nucleus:	one proton	6 protons, 6 neutrons
Electrons:	one	6 at two energy levels
Atomic number:	1	6
Atomic mass:	1	12

Each electron has a single negative charge, each proton has a single positive charge, and neutrons have no charge. The **atomic number** (the lower small number before the symbol) gives the number of protons in the nucleus, which characterises each element and is the same as the number of electrons. The **atomic mass** (the upper small number before the symbol) is the total number of protons plus neutrons in the nucleus.

Electrons exist in one of a series of energy levels, defined in quantum terms by their angular momentum and spin, held to the nucleus because of their different electrical charge. Chemical reactions are to do with the attractions between atoms via their electrons, i.e. electromagnetic forces. The nucleus is held together by nuclear forces, that outweigh the electrical repulsion from having positive charges together. Nuclear physics is about interactions within the nucleus: what makes it have the number of protons and neutrons that it does, and how this might change. The basic types of nuclear reaction occurring are radioactive decay, nuclear fission and nuclear fusion.

The typical size of a nucleus is 10^{-14} m, and of the whole atom, 10^{-10} m. So most of an atom is empty space: if an atom had a nucleus the size of an orange (10 cm), the electrons would be about 1 km away: 10,000 times the nuclear diameter.

Atomic mass and energy

Atomic masses are measured in **atomic mass units**, symbol **amu** or just **u**. One atomic mass unit is defined as one-twelfth the mass of the carbon-12 nucleus, and is equal to 1.66×10^{-27} kg. Originally, an atomic mass unit was intended to be the mass of one proton or one neutron – it was later discovered that these are not exactly equal, and furthermore appear to depend on which element or isotope they are part of – hence its new more precise definition.

Because mass and energy are equivalent ($E = mc^2$, where c is the speed of light, 3×10^8 m s^{-1}), this mass can be expressed in energy units. This was one of Einstein's great insights, part of his formulation of general relativity, that links space and time into four-dimensional curved space-time, where the speed of light is a unifying constant.

To convert from mass to energy in Joules, the energy in 1 amu is given by:

$$E = mc^2 \tag{7.1}$$
$$= 1.66 \times 10^{-27} \times (3 \times 10^8)^2$$
$$= 1.5 \times 10^{-10} \text{ J}$$

The commonly used energy unit for the small amount of energy in a single nucleus is MeV or mega-electron-volts, where 1 amu = 931 MeV. One electron-volt (eV) is the energy gained by a single electron accelerated in an electric field of one volt, given by:

$$E = qV \tag{7.2}$$

Here q, the charge on an electron, is 1.6×10^{-19} coulombs and the voltage V is 1 volt, hence the energy in 1 eV is equal to 1.6×10^{-19} J – a tiny amount of energy. Electron-volts are used to measure energy levels of electrons in atoms.

Neutrons and protons have mass of approximately 1 amu, but not exactly. Their masses are:

$$m_{proton} = 1.007277 \text{ amu}$$
$$m_{neutron} = 1.008665 \text{ amu}$$

The mass of an electron is 0.000549 amu, which is tiny. If an electron had the mass of an orange (say 100 g), then protons would weigh 180 kg, i.e. seven large sacks of potatoes.

Isotopes

Each element has a different atomic number, for instance carbon always has six protons, oxygen always has eight and so on. The atomic number is important as it defines what element the atom actually is and determines its chemical properties. However the number of neutrons in atoms of the same element may vary, giving rise to different **isotopes** with different atomic mass. For instance, carbon has two common isotopes: carbon-12 and carbon-14. Carbon-12, the everyday form, is as shown in Figure 7.1 with six protons and six neutrons, while carbon-14 has six protons and eight neutrons, and is radioactive and unstable. So two isotopes are varieties of the same element with the same number of protons, but with a different number of neutrons.

As isotopes have the same atomic number, they are the same element with identical chemical properties. Carbon-14 can, and does, form exactly the same organic molecules as carbon-12, and is present in all living things. As it reacts chemically in the same way it cannot be separated by chemical means. The only differences are the atomic weight, and that carbon-14 is unstable and so undergoes radioactive decay.

Most elements have more than one isotope, but some are more stable, and therefore more common, than others. The terms **radioisotope** and **radionuclide** are used to denote unstable, radioactive isotopes.

Different isotopes can be identified by means of mass spectrometry. A gaseous stream of the ionised atoms is emitted in an apparatus containing electric or

---

Box 7.1

Sourcing lead pollution by isotopic analysis

Isotopic analysis can be used to identify the source of some environmental pollutants. For instance, naturally occurring lead contains two isotopes, ^{206}Pb and ^{207}Pb. The ratio of the two isotopes depends upon geological conditions of the ore body – principally the age at which it was laid down – and varies for different locations and lead mines. In the UK, historically lead was produced indigenously from many areas in the north of England and Scotland, and is also present in small quantities in coal, with a ratio of $^{206}Pb/^{207}Pb$ of 1.17 to 1.18. By contrast, the lead used as a petrol additive for vehicles in the UK came exclusively from Australian sources, namely the huge mining areas surrounding Broken Hill in New South Wales. This ore body has less ^{206}Pb with a $^{206}Pb/^{207}Pb$ ratio of around 1.04. Studies of lead in dust or in sediments can identify the source of the lead as being from petrol or from coal burning and other sources such as crumbling Victorian paint by measuring this ratio. Farmer *et al.* (1999) show that for Scottish loch sediments, the predominant form of lead in sediments has changed over time from before 1820 to the present day, with the rise of vehicle use and the later introduction of unleaded petrol.

magnetic fields. The atoms will be deflected due to their electric charge, the same for any isotope. The amount of deflection will be dependent upon their momentum, which depends on mass. Hence the different isotopes with different atomic mass will be deflected by different amounts. This means can be used to find the ratios of different isotopes present in a sample of one element.

Isotopic analysis can also be used to determine the source of radioactive pollutants found in soils or waters. Sources including nearby nuclear reactors, fallout from Chernobyl and from atomic weapons testing can be differentiated from a knowledge of the isotopic ratios of the material released in each case.

Binding energy and mass defects

When protons, neutrons and electrons are bound together in an atom, it would take some energy to separate them totally. This energy is the **binding energy**. If the mass of an atom is compared with the sum of the masses of its component parts, the whole is found to be less than the sum of the parts. This difference is the binding energy of the nucleus, also known as the mass defect – mass and energy being equivalent by $E = mc^2$.

Binding energy = Mass of separate particles − Mass of atom

For instance, for helium (^{4}He), the mass of the atom is found to be 4.002603 amu. Helium consists of two neutrons and two protons, being circled by two electrons.

$$\text{Binding energy} = (2 \times 1.007277) + (2 \times 1.008665) + (2 \times 0.000549)$$

$$- 4.002603$$

$$= 4.032982 - 4.002603$$

$$= 0.0304 \text{ amu}$$

This can be expressed as 28.28 MeV, or 4.5×10^{-12} J.

Binding energy is at the heart of all nuclear reactions. Any nucleus has a tendency to occupy the lowest energy state it can – like a ball rolling downhill tends to reach the lowest potential energy state possible, i.e. the bottom of the hill. An unstable isotope is one that is in a high energy state and wants to get rid of some. The way it does this is by nuclear decay – emitting radiation and turning into a different nucleus in a lower energy state.

An analogy is kicking around loads of footballs on a pitch at the top of a hill surrounded by a low wall. As the balls are kicked around, every now and then one will go over the wall and roll away down the hill. It loses energy as it does this, and will never spontaneously roll up the hill, over the wall and on to the pitch – so you end up with fewer and fewer balls. But it needs some energy (from being kicked) to get over the wall, so all the balls don't roll away immediately.

In an unstable isotope, the particles are constantly vibrating; sooner or later one goes over the 'wall' of energy that holds it together and it decays to a lower energy state in the form of a more stable isotope. The reaction can never spontaneously go the other way.

The energy released in the process is the energy used to create heat in a nuclear reactor core, or to heat up the core of the Earth, or to make the Sun shine, or to blow up Hiroshima. Although a single nucleus does not release much energy, because of the huge numbers of nuclei in even 1 g of matter the total amount of energy released can be huge.

Why are some nuclei stable and others not? In a nucleus, the protons repel one another because of their electrical charge, but both protons and neutrons attract by nuclear forces, which reduce with distance. A nucleus could not just contain protons because the electrical force would outweigh the nuclear force. The inclusion of neutrons increases the nuclear forces, holding the nucleus together. Light, stable nuclei such as carbon-12 contain equal numbers of protons and neutrons, which gives them the maximum binding energy; heavier nuclei need more and more neutrons to be stable, thus uranium-238 has 92 protons to 146 neutrons. There are no naturally occurring nuclei heavier than uranium, as they would be unstable and decay into something lighter. A nucleus with too many neutrons, such as carbon-14, will also be unstable, having less binding energy per nucleon than carbon-12.

Types of ionising radiation

Radiation can be used to mean a wide range of things. Here we mean **ionising radiation:** radiation that because of its high energy can cause ionisation – knocking electrons off other atoms or molecules, leaving them charged and reactive. This varies in its physical attributes and in its degree of penetration into solid materials, as shown in Table 7.1.

While most radiation in Table 7.1 actually consists of streams of nuclear particles, γ-rays and X-rays are forms of high energy, high frequency electromagnetic radiation (like light). Because they are of higher energy than light they may ionise other atoms. If the γ-ray interacts with any atom, it can transfer its energy to an electron through resonance with the electron's movement, in a similar way to how microwaves react with water to heat it up in a microwave oven. This gives the electron enough energy to fly off freely if this is more than the energy binding the electron to the nucleus. The amount of energy needed to free an electron entirely from an atom is known as its **ionisation potential**.

Planck's law (Equation (3.15)), shows that the energy carried by one photon is given by $E = hf$, where Planck's constant $h = 6.626 \times 10^{-34}$ J. Hence the higher the frequency, the more energy carried. X-rays have frequencies from around 10^{17} to

Table 7.1 *Types of ionising radiation*

Type	Description	Penetration	Shielding required	Charge
α	Helium nucleus: 2 protons + 2 neutrons	Low	Skin	Positive
β	Electrons	Medium	1–5 mm of metal	Negative
γ-ray, X-ray	High energy electro-magnetic radiation	High	10 mm+ lead or 1 m+ concrete	None
Neutrons	May be fast or slow	High	Several metres of concrete	None
Protons		Medium	Metal foil	Positive
Heavy ions	Variety of decay products	Low	Skin	Yes

Note: radioactive materials may emit any of the above during nuclear reactions and decay.

10^{20} Hz; γ-rays from 10^{20} to 10^{22} Hz – compared with visible light at 10^{15} Hz. For a γ-ray with frequency of 10^{20} Hz, the energy carried will be in the region of 6×10^{-14} J. As an illustration, helium has ionisation potential of 24.6 eV, which is equal to 3.9×10^{-18} J (as 1 eV$=1.6 \times 10^{-19}$ J). In other words the γ-ray has about 10,000 times more energy – ample to ionise the helium atom.

Other types of radiation are similar in that they can cause ionisation. Once an atom is ionised it will be chemically highly reactive – thus ionising radiation causes chemical changes in material it encounters, breaking up stable molecules and creating new ones. In addition neutrons may cause neutron activation. This is when a neutron collides with the nucleus of another atom and reacts to combine with it, turning it into a radioactive isotope of itself, which may be unstable. This can never happen by chemical means alone.

In general, the heavier the particle is the less penetrating it is, as it would need far more kinetic energy to get through anything solid without being stopped. Neutrons however are highly penetrating as they are uncharged, and so are not affected by the electromagnetic fields around nuclei that they pass. The less penetrating forms of radiation are more easily shielded as shown in Table 7.1. However α-particles and heavy ions are still extremely damaging if ingested, by swallowing or breathing in as dust, allowing access to your inner organs.

Units of radiation measurement

Radioactivity is measured in terms of the number of radioactive decays that take place per second. One decay per second is one Becquerel (Bq). So if monitoring shows that a moor in Cumbria has radioactivity of 17,000 Bq per square metre, that means that on average, in 1 m^2 of moorland in one second there will be 17,000 radioactive nuclei decaying.

When looking at the dose that someone has received, the important factor is how much energy is released in the body by the radiation reaching it. Dose rates are measured in Grays (Gy), where 1 Gray means releasing energy of 1 J kg^{-1} bodyweight of the exposed part. The dose may therefore be higher for a child because of their lower bodyweight, than for an adult exposed to the same level of radiation – this reflects the fact that they are more likely to suffer ill effects.

In biological terms, some types of radioactivity are more harmful than others. So for radioactivity safety standards, the **dose equivalent** is measured, for which the unit used is the Sievert (Sv). A sievert is the dose received multiplied by the **quality factor** (QF), as follows:

X-rays, γ-rays, β	QF = 1	Least damaging
Thermal neutrons	QF = 2	↓
Fast neutrons, protons	QF = 10	↓
Heavy ions	QF = 20	Most damaging

So for a given energy emitted, heavy ions would be 20 times more damaging than X-rays, and hence the dose equivalent in Sievert would be 20 times higher than the same dose of X-rays in Grays. Exposure limits are set in Sieverts as the most appropriate measure of potential damage.

Table 7.2 summarises the units used for radioactivity. The old units of Curie, rad and rem are still commonly in use and so are also shown in the table.

Conversion from Becquerels to Grays or Sieverts is not straightforward, as the former is a measure of number of reactions and the latter two are measures of energy. The conversion depends upon the type of radiation emitted and how much energy the reactions release.

Table 7.2 *Units of radioactivity measurement*

Quantity	SI unit	Old unit
Radioactivity	Becquerel 1 Bq = 1decay per second	Curie 1 Ci = 3.7 × 10^{10} Bq
Dose	Gray 1 Gy = 1 J kg^{-1}	Rad 1 rad = 0.01 Gy
Dose equivalent	Sievert 1 Sv = 1 Gray × QF	Rem 1 rem = 1 rad × QF

Radioactive decay

Some isotopes have a nucleus that will decay spontaneously into another element, referred to as the daughter product. They may do this by emission of either α- or β-particles. For instance, natural uranium-238 (^{238}U) decays into thorium (Th) emitting an α-particle, which is the same thing as a helium nucleus (He):

$$^{238}_{92}\text{U} \rightarrow \, ^{234}_{90}\text{Th} + \, ^{4}_{2}\text{He} \tag{1}$$

The reaction may either be written as above, or with α instead of the He symbol.

In any nuclear reaction, the sum of atomic masses (i.e. the number of nuclear particles, neutrons plus protons) and the sum of atomic numbers (the number of protons, or nuclear charge) on each side is the same due to the fundamental conservation laws. This principle is useful in predicting how things will decay. In reaction (1), uranium-238 has 92 protons and 146 neutrons, giving a total atomic mass of 238. Thorium has 90 protons and 144 neutrons, while helium has two protons and two neutrons. So overall, the total atomic mass stays constant (238 = 234 + 4), as does the atomic number (92 = 90 + 2).

Energy will be released in this process, in the form of increased thermal energy or heat. This energy comes from the binding energy of the nuclei. If the mass of each were calculated, the sum of the masses of thorium and helium would be found to be less than that of uranium. So the uranium decays to reach this lower energy state, and the difference is released as heat. To make the process go the other way, this energy would need to be supplied.

A typical β-decay is that undergone by the radioactive isotope of carbon, carbon-14, which decays to form nitrogen:

$$^{14}_{6}\text{C} \rightarrow \, ^{14}_{7}\text{N} + \, ^{0}_{-1}\text{e} \tag{2}$$

Again this could be written with either e for electron, or β for β-particle as the two are identical. You may wonder where the electron comes from in this reaction, as the nucleus does not contain electrons. In fact in β-decay, a neutron in the nucleus itself splits up to form a proton and an electron:

$$\text{n} \rightarrow \text{p} + \text{e}$$

Overall the charge remains zero and the atomic number equal to 1, although there is a small loss in mass which is released as energy.

Radioactive decay chains

Once a decay reaction has occurred, it may produce an element that is still radioactive, and which will decay into something else. For instance, uranium decays into thorium, which can then decay into protactinium (Pa), then into a new isotope of uranium, and so on until a stable isotope is formed, generally lead (^{206}Pb).

$$^{238}_{92}\text{U} \xrightarrow{\alpha} {}^{234}_{90}\text{Th} \xrightarrow{\beta} {}^{234}_{91}\text{Pa} \xrightarrow{\beta} {}^{234}_{92}\text{U} \xrightarrow{\alpha} {}^{230}_{90}\text{Th} \rightarrow {}^{226}_{88}\text{Ra} \rightarrow {}^{222}_{86}\text{Rn} \rightarrow$$

$$\rightarrow {}^{218}_{84}\text{Po} \rightarrow {}^{214}_{82}\text{Pb} \rightarrow {}^{214}_{83}\text{Bi} \rightarrow {}^{214}_{84}\text{Po} \rightarrow {}^{210}_{82}\text{Pb} \rightarrow {}^{210}_{83}\text{Bi} \rightarrow {}^{210}_{84}\text{Po} \rightarrow {}^{206}_{82}\text{Pb} \qquad (3)$$

At each stage, α or β radiation is emitted, as shown for the first four steps. You can work out which, as each time either the atomic mass decreases by 4 and atomic number by 2 (α decay), or the mass stays the same and the number increases by 1 (β decay), to balance each reaction. Each subsequent radionuclide will thus have an atomic mass lower or the same, but the atomic number may increase or decrease. The chain continues, producing smaller nuclei, until a stable isotope forms. At any time, there will be a mix of all of these daughter products in varying proportions, that can be calculated knowing the half-life for each isotope. Eventually, everything will turn to stable lead-206, but that will take billions of years.

Decay rates and half-lives

The rate at which a radioactive substance decays is proportional to the amount of it remaining. This leads to an exponential decay law. It can be described by an element's **half-life**. The half-life is the time taken for half of the isotope to decay, also the time for activity to reduce by half.

For instance the half-life of caesium-137 is 30.3 years. If a nuclear power station produces waste containing caesium with activity of 8×10^{18} Bq, after 30.3 years, half of it would have decayed (turning into another element) reducing activity to

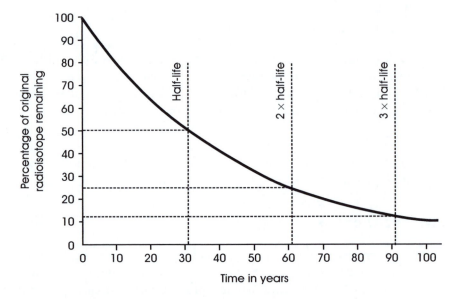

Figure 7.2 *Radioactive decay of Caesium-137, half-life 30.3 years.*

4×10^{18} Bq. After a further 30.3 years, half of the remaining amount would decay, leaving 2×10^{18} Bq. After another 30.3 years, there would be 1×10^{18} Bq produced, and so on. This is illustrated in Figure 7.2.

The graph is an exponential decay curve, with an equation of the form

$$N = N_0 \, e^{(-\lambda t)} \tag{7.3}$$

where N_0 and N are the number of nuclei at the start and after time t, and λ is a decay constant. To express this in terms of the half-life $t_{1/2}$, enter the figures for $t = t_{1/2}$, when $N/N_0 = 0.5$:

$$e^{-\lambda t_{1/2}} = 0.5$$

$$\text{Take logs:} \quad \lambda = 0.693/t_{1/2}$$

$$\text{This gives:} \quad N = N_0 \, e^{(-0.693 t/t_{1/2})} \tag{7.4}$$

To work out the time taken to decay, this equation can be rearranged to:

$$t = \frac{t_{1/2} \ln (N/N_0)}{-0.693} \tag{7.5}$$

Half-lives of different isotopes vary enormously – from seconds to billions of years.

Table 7.3 *Half-lives of various isotopes*

Isotope	Half-life
Nitrogen-16	7 sec
Argon-41	1.8 hours
Radon-222	3.8 days
Iodine-131	8 days
Strontium-90	29 years
Radium-226	1,599 years
Plutonium-239	24,000 years
Uranium-235	7×10^8 years
Uranium-238	4.5×10^9 years

Half-lives are related to the probability of any one nucleus decaying, as decays that are most likely to occur will result in short half-lives. They depend upon the energy levels and strength or forces within the nucleus, with the most unstable nuclei having the shortest half-lives. These are by definition the most radioactive isotopes, so in the decay of a mix of radioisotopes, such as in nuclear waste, there is a natural progression from predominantly highly active, short half-life isotopes to less active but very long-lived isotopes. However long-lived isotopes may decay into shorter-lived ones, and vice versa.

The isotopes of most concern environmentally are those with long half-lives, as these may accumulate in the environment, and may require containment over very long periods. Half-lives cannot be altered by any physical or chemical means, so this process cannot be speeded up.

Carbon dating and other radiometric dating techniques

Carbon dating uses a measurement of the proportion of carbon's radioactive isotope to calculate the age of ancient organic objects.

Carbon-14 (^{14}C) has a half-life of 5,730 ±40 years. It decays according to reaction (2) into nitrogen, emitting a β-particle. After 5,730 years, half the ^{14}C will have decayed. ^{14}C exists in the atmosphere as it is created by cosmic rays striking the ionosphere. So a certain fixed proportion of all CO_2 in the atmosphere contains ^{14}C rather than stable ^{12}C. When a plant grows, it takes up ^{14}C in this fixed proportion. The ^{14}C then decays back into nitrogen over the next few thousand years.

In carbon dating, the proportion of ^{14}C to ^{12}C is measured, which can alternatively be deduced from the decay rate of the sample. This is compared with the fixed proportion in the atmosphere. A proportion of half of that in the atmosphere indicates an age of 5,730 years (the half life of ^{14}C), while a quarter would be 11,460 years and so on. In general Equation (7.5) above will be needed to work out the age.

By this method, organic materials can be dated, such as ancient manuscripts, historical artefacts made from wood or leather, pollen grains, or bones. This method is useful for objects from a few hundred to around 40,000 years old, and for anything containing organic material. Accuracy depends upon the age, the nature and condition of the sample but can often be within 5 per cent. Some discrepancies exist due to variations in the ^{14}C/^{12}C ratio in the atmosphere. These can be natural, because of variations in the Earth's magnetic field and cosmic rays, and changes in oceanic uptake rates. More recently, emissions of CO_2 from burning fossil fuel have reduced the atmospheric ^{14}C proportion by releasing carbon that was laid down

---

Box 7.2

Palaeoecology and peat bogs

One interesting application of carbon dating is in palaeoecology. Historical ecosystems can be studied from their remnants, preserved in peat bogs.

In cores taken from a bog, the different layers correspond to increasingly ancient times as the peat has been laid down. From each layer, ancient pollen grains can be extracted, which are still preserved well enough to be identifiable. The range of species found is found to change with depth. Using carbon dating on the pollen and the surrounding peat, the layers can be dated quite accurately, giving a 'calendar' of the commoner species' occurrence over thousands of years.

The pollen record shows ancient forests flourishing and changing species as newly evolved varieties took over. Some species vanish altogether as they die out or the climate changes; others appear or increase in importance. Then around 6000 BC in the UK a major change starts to happen – tree pollen gets rarer, while various grasses and cereals take over. This is the start of settled human agriculture, when the forests started to lose their dominance and mankind shaped the environment. Over the following thousands of years the record continues to trace the introduction of new species and development of crops, with forests becoming less and less important, until our modern day landscape of monoculture and deforested moorland emerges.

Table 7.4 *Isotopes used for radiometric dating*

Parent isotope	Daughter isotope	Half-life (years)
Uranium-235	Lead-207	704 million
Uranium-238	Lead-206	4.5 billion
Thorium-232	Lead-208	14 billion
Potassium-40	Argon-40	1.25 billion
Rubidium-87	Strontium-87	48.8 billion

millions of years ago and so is depleted in ^{14}C. Conversely, atomic tests and nuclear explosions have increased ^{14}C by releasing radioactive material into the atmosphere.

A number of historically important artefacts have been carbon dated using this method. For instance, charcoal from the Stonehenge site has been dated at 1846 BC ± 275 years, consistent with other evidence of the age of the structure.

For older objects and for rocks and minerals other elements/isotopes can be used, in a process known generally as radiometric dating. Isotopes commonly used in dating rocks include those shown in Table 7.4. In these cases, the principle is slightly different to carbon dating. Unlike carbon, these elements are not associated with organic mater or in exchange with the atmosphere. The date is estimated by measuring the proportions of the parent and the daughter material (usually lead). The long half-lives of these isotopes make them suitable for measuring ages of rocks, measured in millions or billions of years, unlike carbon-14 which has a much shorter half-life.

Radiometric dating is limited to rock containing the appropriate isotope, where the parent isotope was included during the rock's formation process but the daughter isotope was not (due to chemical processes in sedimentation, or crystallisation of igneous rocks). As the isotope decays, the daughter isotope is incorporated into the crystals of the rock. The proportion of daughter to parent may then be used to estimate the date of formation, which may also include dates of events such as volcanic lava flows or major meteorite strikes. Adjustments may need to be made if some of the daughter material was believed to be present at formation, or has since escaped from the rock due to metamorphic events, leaching or diffusion since formation. In some cases the same daughter product may be produced from two or more parent isotopes. These problems may result in inaccurate dating from these methods.

The details vary with each isotope used, but these radiometric dating techniques are limited to finding the date of formation, or in some cases of metamorphosis, of the rock, and to rocks that are at least tens of millions of years old. They could not be used to find, for instance, the date of manufacture of stone tools, or the date of a recent volcanic event. Many rocks cannot be dated by radiometric measurements, if they do not contain the appropriate isotopes.

Radiometric dating techniques are important in dating of ice-cores, used in studying climate change and atmospheric composition over very long periods.

Radioactive decay of gases dissolved in the ice can be used to estimate the age from isotopic ratios.

Biological impacts of ionising radiation

By its definition ionising radiation has enough energy to ionise atoms of any substance it encounters. If radiation enters the body, it can cause chemical damage to cells or to DNA, which can cause cancer. In large doses, **radiation sickness** is produced, caused by large-scale cell damage leading to sickness and sometimes death within days or weeks.

In the longer term, damage can include both production of many different cancers, including leukaemia and genetic mutation affecting future children of the exposed person. Cancers may appear years after exposure – the long delay is partly because radioactive material can become lodged in the body, where it stays for many years gradually decaying and releasing energy and β- or α-radiation which does the damage. Where organs such as the liver concentrate particular elements, specific problems can occur. Thyroid cancer produced by iodine-131 ingestion is a particular concern as iodine is absorbed and concentrated by the thyroid gland. The risk can be reduced by consumption of iodine tablets containing the stable isotope, which is accumulated in the thyroid causing the radioiodine to be excreted, but only if consumed before significant exposure.

Radiation doses and dose limits

We are exposed to radiation from a number of sources, of which artificial sources are a small fraction. Table 7.5 shows estimates of the average exposure of individuals in the UK from the National Radiological Protection Board (NRPB). The current dose limit set by the International Committee on Radiological Protection (ICRP) for exposure to the public for all artificial sources of radiation (excluding medical) is 1 mSv per year, averaged over five years. Occupational dose limits are higher, with a dose limit for radiation workers of 20 mSv per year averaged over five years, and regulations on the maximum number of hours that a worker can be exposed to higher levels.

These averages will hide large individual variations. As aircraft fly in the stratosphere passengers and crew are subject to significantly higher levels of cosmic radiation. Radon gas occurs naturally in certain rocks, notably granites, and can accumulate in houses under certain conditions (Box 7.3). This can produce high doses among those affected. We are also subject to naturally occurring gamma radiation from the Earth, and a small amount of radioactive material is present in all our bodies in the form of natural isotopes of potassium and carbon, ^{40}K and ^{14}C.

Table 7.5 *Average radiation doses to British public*

Source	Dose, mSv
Natural sources	
Cosmic	0.25
Terrestrial gamma	0.35
Internal radionuclides	0.3
Radon-222	1.3
Thoron (Radon-220)	0.1
Artificial sources	
Medical	0.3
Occupational	0.005
Weapons fallout	0.01
Discharges	<0.001
Consumer products	<0.001
Total	2.61

Source: Green *et al.* (1992)

These sources comprise the background radiation, and are normally higher than any exposure to artificially produced radiation.

Of the artificial sources, exposure from medical and dental X-rays delivers a large dose to the affected patient, although the health benefits presumably outweigh the costs. Exposure of those working in the nuclear industry, coal miners, X-ray technicians and certain other occupations will be higher than of the public, but is strictly regulated by government exposure limits and monitoring. Consumer products contribute, including digital watches containing lithium, and the atmospheric atomic weapons tests of the 1950s and 1960s produced fallout globally that we still experience. In addition there are discharges from licensed nuclear installations, considered on pp. 269–70.

Environmental pathways of radioisotopes

Consideration of the pathways by which radiation can enter our bodies reveals factors that create greater variations in individual doses. While gamma rays and neutrons can enter directly by exposure to radiation, other forms cannot penetrate the skin. Gases such as radon and dust may be inhaled, and soluble material taken in from water. Radioactive material may enter the food chain and be eaten in meat, milk, fish or vegetable matter, or as dust or dirt consumed with food, particularly by children. The pathways through the environment may be modelled to estimate doses – atmospheric transport and deposition, groundwater movements, leaching and water transport of these pollutants, which may be followed by their biological uptake into the foodchain, as illustrated in Figure 7.3.

The pathways available to a particular radionuclide depend upon its physical and chemical properties, discussed in detail by Kathren (1984: 271–275). Of particular concern are radioisotopes that may mimic important nutrients. For instance, strontium-90 has similar chemical properties to calcium. It can be taken up into milk via the diet of a cow, and then when drunk absorbed by the body as a calcium substitute and incorporated into bones and teeth, particularly in children.

While many environmental processes act to dilute and disperse pollutants, in some cases natural processes can lead to concentration and bioaccumulation. Once in the soil or water, many radionuclides will not be taken up by plants or seaweed to any

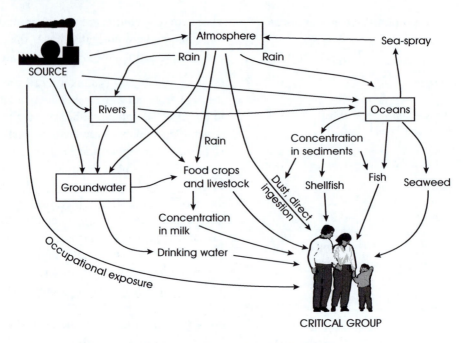

Figure 7.3 *Environmental pathways for uptake of radioisotopes.*

great extent, but some will be selectively absorbed leading to high concentrations in the plant. Ocean currents may carry concentrated pollutant bodies some distance before dispersal occurs, and sedimentation processes can in some cases lead to concentration. Shellfish and marine plants such as seaweeds can selectively uptake certain nutrients, such as iodine and phosphorus. If this seaweed is eaten either directly or by fish that then are consumed by humans, large doses may be received.

The amounts of any particular radionuclide reaching humans will be greater by some pathways than others, due to the many possible routes it could take. The pathway that provides the greatest dose to a particular population is termed the critical pathway. The critical population most at risk due to this pathway may be defined by their location and/or diet, such as those eating laver bread made from seaweed from the Irish Sea in Wales, breast-fed babies or children drinking milk produced close to Chernobyl. In each case the dose to these critical populations can be estimated, including the dose to critical organs such as the thyroid, and steps taken to minimise it to within safe levels.

Risk analysis

Many of the impacts of low doses of radiation are **stochastic** effects, expressed in terms of the probabilistic risk. For instance, someone exposed to a dose of 100 mSv has an excess risk of 1 in 200 of contracting a fatal cancer, while a dose of 1 mSv carries a risk of 1 in 20,000. In other words, if 200 people are exposed to 100 mSv, on average one more will die of cancer than if unexposed. However as cancer is a common cause of death, it might be expected that 40 of these people would die of cancer anyway, if not exposed; 41 will die if exposed.

If a large number of people are exposed to a small dose, the total **expected value** of excess deaths may be calculated, by multiplying the risk factor by the population exposed. So if one million people were exposed to 1 mSv (with risk factor of 1 in 200,000 given above), the expected increase in cancer deaths would be 5. The total deaths from cancer might be 200,000 if unexposed, 200,005 if exposed.

It is evident that these effects may be statistically difficult to measure, in particular in the case of very small doses affecting very large populations (such as after Chernobyl) despite being real for the unlucky victims. Research on the impacts of radiation on human health is limited, as it is not ethically possible to produce experimental evidence, apart from on animals. Much of our knowledge comes from epidemiological studies of Japanese survivors of the atomic bomb, and from military personnel exposed to radiation in atomic weapons tests. Given these limited sources of data, the effect of low doses over high populations and long time-scales must be extrapolated, so all figures are estimates.

Risk assessment may be extended to nuclear safety in estimating the risk of a particular accident occurring. Any accidental release of radiation will be the result of a chain of incidents – system failure, operator error, failure of safety systems and so forth. Each of these individual incidents has a certain probability of occurring, many of which can be estimated from experience of reactor operation. These estimates can then be combined through the chain to estimate the probability of the release of radiation, or a major accident. An understanding of the dispersion of radiation, the populations exposed and the impact of the dose they receive can then be used to calculate the possible consequences of an accident. The most serious accidents have very low probabilities, for instance the probability of a major incident causing 1,000 deaths has been estimated at around 1 in 100 million per reactor year (Scott and Johnson 1997). These estimates are only as good as the procedure used to calculate them. If the analysis does not allow sufficiently for operator error, or overlooks an important factor such as metal fatigue, or misses one possible event sequence that could cause a release, or assumes safety standards in developing countries are as high as those in the west, the probability could be higher.

The estimated risks from radiation, whether from routine or accidental releases, are far lower than many other risks we are exposed to. Natural disasters such as earthquakes and tornadoes, and artificial risks such as the uninvited effects on health of passive smoking, traffic accidents or the occupational health risks suffered by coal miners account for many more deaths than radiation. Whether we accept the estimates as accurate, and whether these small increased risks are justified by the benefits of nuclear power production, is a controversial question.

Box 7.3

Radon gas in homes

Radon-222 is one of the natural decay products of uranium-238 which occurs in some rocks. As radon is a gas, when it is generated it can move readily through pore spaces in the rocks to the surface. Outside, it will then rapidly disperse in the atmosphere, and decays with a half life of 3.8 days. However radon can build up indoors to dangerous levels (Green *et al.* 1992). Radon enters buildings from the ground below them and also from building materials, with some contribution from external air. Radon from air in the ground is drawn into buildings because of lower pressure inside, as a result of heating of the house and air being drawn out through chimneys and extractor fans.

The concentration of radon in homes depends principally on the underlying rock type, as higher levels of uranium are present in some granites, Old Red Sandstone and limestones, together with sufficiently high permeability for radon to be released at the surface. For this reason the vast majority of dwellings with problem levels in the UK are found in Cornwall and Devon, together with some in Derbyshire and north Yorkshire.

Damage is caused principally by inhalation of daughter products of radon, which then decay producing alpha-radiation. This damages cells in the lungs leading to lung cancer. Radon exposure at levels commonly found in homes poses significant risks of lung cancer, producing an estimated 2,500 extra cases in the UK annually. The average level for homes in the UK is 21 Bq m^{-3}, which would be expected to produce a lifetime risk of 0.3 per cent. Given normal incidence of lung cancer of 6 per cent, this implies that 1 in 20 cases are caused by radon exposure. However for certain areas typical concentrations are much higher, with 100–400 Bq m^{-3} common in Cornwall. Lifetime risk of lung cancer increases to 6 per cent at 400 Bq m^{-3}, doubling the risk of the disease to those exposed. In the extreme some households have been identified with concentrations up to 10,000 Bq m^{-3}, implying a dose of 500 mSv to occupants, which reaches high dose levels and may have short-term, non-stochastic ill effects.

It is evident that indoor radon pollution presents significant risks, orders of magnitude higher than the risks from routine discharges from nuclear power for instance. The main actions that should be taken are monitoring of affected areas and implementing preventive measures where necessary, which including sealing of floors to prevent radon entry, and ventilation of sub-floors.

Power from nuclear fission

All nuclear power stations work upon the principle of nuclear fission. In a nuclear fission, rather than a nucleus spontaneously decaying by α or β radiation, it is split into two smaller daughter nuclei. Fission is made to occur in nuclear power by bombarding a uranium nucleus with neutrons. The ^{235}U nucleus absorbs a neutron, producing ^{236}U in an excited state (the * in the reaction means it is excited, or has extra energy). This is unstable because it has too much energy, and it splits into two fragments or daughter products.

$$^{235}_{92}\text{U} + {}^1_0\text{n} \rightarrow {}^{236}_{92}\text{U*} \rightarrow {}^{90}_{36}\text{Kr} + {}^{144}_{56}\text{Ba} + 2{}^1_0\text{n} \tag{4}$$

In this case, the fission products are krypton and barium. However many different daughter products may be formed, with atomic masses from 75 to 160. The different daughter products will also produce different numbers of neutrons. As these products are radioactive isotopes, they will then decay into something else, releasing more radiation, and are much more radioactive than the original uranium.

If an isotope will undergo a fission reaction of this nature, it is known as a **fissile** isotope. Natural uranium contains mainly two isotopes ^{238}U and ^{235}U. ^{235}U is a fissile isotope that is needed for nuclear reactors, and it comprises only about 0.7 per cent of natural uranium. The most common isotope, ^{238}U, is not fissile. Another isotope of uranium, ^{233}U, is also fissile, which is produced in reactor cores. Plutonium is the only other fissile isotope commonly encountered, which does not occur naturally but is produced in nuclear reactors.

The fission reaction will occur with a certain probability that depends upon the energy of the neutron it reacts with, and the number of neutrons. Fission is most likely to occur with neutrons of fairly low energy (termed slow or thermal neutrons), and if they are present in large numbers. Controlling the neutron density and their energy is therefore the key to controlling the reaction rate.

Energy released by nuclear fission

The energy for nuclear power comes from the binding energy of the nuclei involved. If the total mass on each side of a nuclear reaction is calculated, it will be found that a bit has 'gone missing'. This bit has been converted into energy (according to Equation (7.1), $E = mc^2$), which is released as heat.

For instance, one possible fission reaction taking place in a reactor is given by:

$$^{235}_{92}\text{U} + {}^1_0\text{n} \rightarrow {}^{148}_{57}\text{La} + {}^{85}_{35}\text{Br} + 3{}^1_0\text{n} \tag{5}$$

Uranium-235 absorbs a neutron, then splits to form lanthanum, bromine and three neutrons. The atomic masses in atomic mass units (amu) are as follows:

Uranium-235	235.1
Lanthanum-148	148.0
Bromine-85	84.9
Neutron	1.009

So the total mass on each side is (in atomic mass units):

$235.1 + 1.009 \quad = 148.0 + 84.9 + (3 \times 1.009) + \text{energy released}$

$236.109 \qquad\quad = 235.927 + \text{energy released}$

$\text{energy released} = 0.182 \text{ amu}$

Given that 1 amu $= 1.5 \times 10^{-10}$ J, and 1 kg uranium contains 26×10^{23} atoms the energy released from the fission of 1 kg uranium-235 would be given by:

$$E = 0.182 \times 1.5 \times 10^{-10} \text{ J} \times 26 \times 10^{23}$$

$$= 7.1 \times 10^{13} \text{ J}$$

$$= 71 \times 10^{6} \text{ MJ}$$

As natural uranium contains 0.7 per cent ^{235}U, the energy present in 1 kg of natural uranium is about 5×10^{5} MJ. This compares with about 29 MJ in 1 kg of coal – the energy content is over 17,000 thousand times more in uranium.

The exact amount given off depends on the daughter products, of which there are many, although this energy release is typical of many possible reactions.

Critical mass

In a nuclear power station or a nuclear bomb, the basic reaction is the same. Uranium-235 (or plutonium) forms the basis for a self-propagating chain reaction. The fission reactions in (4) and (5) not only release energy, they also release some excess neutrons. These neutrons can then react with more ^{235}U nuclei, producing more neutrons – the number depends on the daughter products, but the average is 2.5 neutrons per decay. So the reaction can very rapidly expand, with each reaction setting off two or three more, and each producing a large amount of energy – like a chain letter (Figure 7.4).

If there is only a small amount of uranium present, many of the neutrons produced will be lost by flying off away from the uranium or being absorbed by other materials. Past a certain mass however, there will be fewer neutrons escaping than needed to sustain the reaction – so it goes 'critical', suddenly increasing activity. This is the principle behind an atomic bomb: jam two sub-critical lumps of fissile uranium or plutonium together to exceed the critical mass, and stand well back!

Nuclear power – types of fission reactor

Commercial nuclear power is produced by fission in thermal nuclear stations. These are so called because they rely on thermal neutrons, or slow, lower energy neutrons, to propagate the fission reaction. A thermal nuclear power station contains the following components:

Fuel rods contain the fissile material. This may be natural uranium, enriched uranium containing a higher proportion of fissile ^{235}U, or mixed oxide fuel, containing uranium and plutonium oxides.

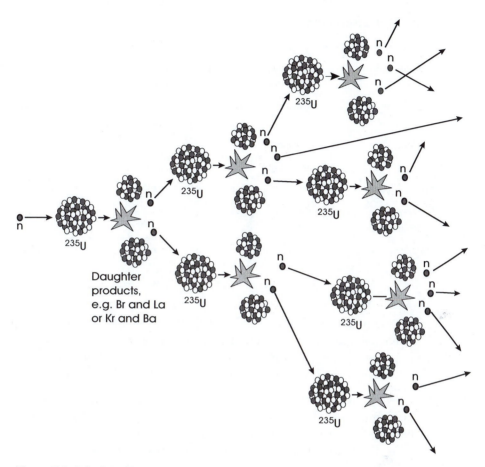

Figure 7.4 *Principle of the chain reaction.*

The **moderator** acts to slow down neutrons, which increases the chance of fission occurring. Slow neutrons will react with another uranium nucleus more readily than fast neutrons.

Control rods are usually made of boron, and work by absorbing neutrons, preventing criticality and controlling the reaction rate.

Coolant. Water or high pressure gas is used to reduce the temperature of the core and transport heat to the boilers, and to prevent overheating of the core.

A **steam turbine** and **generator** converts the heat into electricity, which are essentially identical to those used in a fossil fuel station.

The key components of nuclear power stations are very similar, but the different types of reactor are generally named after the type of coolant, the type of fuel or the type of moderator used.

Table 7.6 *Features of typical reactors*

Reactor type	Magnox	AGR	PWR
Named after	Magnesium oxide in fuel casing	Advanced gas-cooled reactor	Pressurised water reactor
Fuel	Natural U metal	U oxide, 2% enriched	U oxide, 3.2% enriched
Moderator	Graphite	Graphite	Water
Coolant	CO_2 gas	CO_2 gas	Water
Power density	0.57 kW l^{-1}	2.4 kW l^{-1}	105 kW l^{-1}
Coolant temperature	390°C	635°C	325°C
Coolant pressure	8.9 atmos	42 atmos	158 atmos
Efficiency	28%	40%	33%

Other reactor designs in use include the BWR or boiling water reactor which is water cooled; the Canadian CANDU, a heavy water reactor, using water containing deuterium (^2H) as moderator, and the RBMK, the Soviet design used at Chernobyl and elsewhere, which combines water cooling with a graphite moderator.

The Magnox station, designed and built in the UK, was among the earliest designs for nuclear power stations, and due to its low power density and temperature it is fairly inefficient. Magnox stations are now nearing the end of their lives and most are being decommissioned, presenting a major waste disposal issue as discussed on pp. 270–2.

After the Magnox programme, the UK moved on to build AGR stations which now comprise the majority of UK nuclear power plants. These together with PWRs use enriched fuel, increasing the power density. AGRs run at high temperatures which improves efficiency. PWRs were initially designed for use in American nuclear submarines, needing very high power density to operate in a confined space. Adapted to power production, the PWR has now taken over as the most popular design worldwide – they have a relatively good safety and operational record and are more economic to run than most other designs.

Control of nuclear reactors

The aim of nuclear reactor control is to maintain the chain reaction in a critical state, such that each nuclear fission produces enough neutrons to set off exactly one more fission, no more and no less. Under normal conditions, the reactor is controlled by absorbing neutrons in movable control rods, thus slowing down the reaction.

The moderator acts to slow the neutrons to the energy level most suited to the reaction and so maintains the power at a constant rate. The energy produced is released as heat, which is removed by the coolant so that the temperature remains constant.

In a water-cooled reactor such as a PWR the water will act as coolant, moderator (slowing the neutrons and thus increasing the fission rate) and neutron absorber (slowing the reaction). Under normal conditions the operation of the control rods and the moderator are combined with the effect of the water present to maintain a stable neutron density and constant power output. If the water in the core starts to boil producing bubbles of steam within the reactor core (termed **voids**), the number of neutrons absorbed by the water will decrease as its density decreases, tending to increase the reaction rate. However its moderating effect will also decrease, tending to slow the reaction. Whether the overall net effect is to speed or to slow the reaction depends upon the detailed configuration of the reactor. Where the net effect is to increase the reaction, as is the case in the Soviet designed RBMK reactors, the reactor is said to have a **positive void coefficient**. This is a potentially dangerous and unstable situation, as an increase in reactor power will increase temperature, producing more steam and more voids.

One of the keys to controlling a nuclear reactor is the presence of **delayed neutrons**. When a fission reaction occurs, it is virtually instantaneous, and the neutrons produced set off further reactions virtually immediately. This would appear to make it extremely difficult to control the reactor as any very slight increase in reaction rate would rapidly lead to supercriticality and an uncontrollable explosion. Fortunately, some neutrons are delayed by a few seconds. These neutrons come from daughter products that decay producing neutrons with half-lives of a few seconds or minutes. The existence of these delayed neutrons means there is a reasonable length of time in which a small increase in reaction rate can be responded to, reducing activity back to its steady state.

Fast-breeder reactors

In a fast-breeder reactor (FBR), the aim is to use not only the uranium-235 and plutonium as fuel, but to react the uranium-238 that is far more abundant. As natural uranium contains 99.3 per cent ^{238}U and only 0.7 per cent ^{235}U, world uranium reserves are limited to a few decades if used in conventional reactors, using only the ^{235}U (without reprocessing). In fast-breeder reactors, it would last for up to 300 times as long, hence FBRs are seen as important if nuclear power is to increase significantly.

A fast-breeder is so called because it needs fast (high-energy) neutrons to make the reaction occur, and because it breeds its own new fuel from the common isotope ^{238}U. The process is not fast, taking many years to produce significant amounts of plutonium. The ^{238}U absorbs fast neutrons via a series of reactions to produce either plutonium or uranium-233. Both of these are fissile and then produce energy in a fission reaction as in a conventional reactor.

The first of these nuclear reactions is known as the **breeding reaction,** and has the typical form:

$$^{238}_{92}U + ^{1}_{0}n \rightarrow ^{239}_{92}U \rightarrow ^{239}_{93}Np + \beta^- \rightarrow ^{239}_{94}Pu + \beta^- \tag{6}$$

Uranium-238 is transmuted into plutonium, ^{239}Pu. The fast neutrons in a fast-breeder make the reaction most likely to happen, although in a conventional nuclear reactor a small proportion of the ^{238}U will react in this way, which is why the waste fuel contains plutonium. A similar reaction produces ^{233}U.

The breeding reaction does not produce significant amounts of energy – this comes from a fission reaction which is essentially the same as in a conventional reactor, as in ^{235}U reactions (4) and (5). Both ^{233}U and plutonium undergo reactions of this nature, producing a variety of daughter products.

Control and operation of an FBR is more complicated than a conventional reactor, as the breeder reaction (6) needs fast neutrons, while the fission reaction proceeds more rapidly with slow neutrons. More neutrons are needed than in a thermal nuclear reactor, because every fission reaction must produce two neutrons on average – one to spark off another fission, and one to initiate a breeding reaction. A much higher neutron density is therefore needed to keep both reactions going. Because of this, highly enriched fuel is used, containing high levels of ^{233}U or plutonium, despite the fact that (eventually) the ^{238}U will be used – however it can take 20 years before the original amount of fissile material is reproduced by the breeding reaction.

The FBR uses liquid sodium as a coolant. This is used because water or carbon dioxide used in conventional nuclear plants also act as a moderator, slowing down the neutrons, which would prevent the 'breeding' reactions occurring. Sodium has good heat conductivity and does not slow down or absorb as many neutrons. The reactor core is smaller and hotter, with very high power density – the French Super-Phoenix plant has a power density of 277 kW l^{-1}, well over twice that of a PWR. This means that a very effective coolant is required to remove all the heat produced, another reason why liquid sodium is used. FBRs are in some respects more dangerous, and have the disadvantage of being an intrinsic part of the 'plutonium economy', involving close historical links with nuclear arms manufacture.

Fast-breeders are intrinsically more expensive than conventional reactors, because of the use of sodium, higher temperatures and more elaborate safety mechanisms needed to prevent accidents. Production of plutonium is now seen as a problem rather than a lucrative by-product. The result is that worldwide interest in FBRs has waned: the US programme was terminated in 1977 and the only UK FBR, at Dounreay in northern Scotland, closed in 1994. In France, the large FBR called Super-Phoenix operated commercially but has now closed indefinitely principally for cost reasons. There is one remaining FBR in Japan, at Monju, and they have also faced safety problems and had closures.

Nuclear safety and nuclear 'incidents'

Nuclear reactors are designed to have fail-safe systems, so that under any unexplained increase in temperature or reaction rate, the control rods will drop, under gravity if all power has failed, into the core reducing the activity. All radioactive components of a reactor are contained in a sealed system, but containment failure would result in the release of radioactive gas and/or water containing dissolved contaminants to the environment. This could occur through unmonitored cracking or corrosion. Secondary containment is designed to prevent this.

The loss of moderator would result from a water or CO_2 leak in thermal reactors. Neutrons will not be slowed down and will escape, so the reaction will slow and stop. This is therefore a self-resolving situation.

If a reactor were to suffer loss of coolant, such as from a leak in the pressurised water or gas cooling system, the reactor would overheat. This will not by itself cause the reactor to increase in activity, but it may cause damage or buckling to fuel rods and control rods from thermal expansion, and the reactor must therefore be shut down. In a water-cooled reactor with a negative void coefficient such as a PWR, where the coolant acts as moderator, loss of coolant would slow the reaction down. A reactor such as the RBMK with a positive void coefficient is more likely to become super critical under these conditions, as the reduction in neutron absorption predominates (see Box 7.4).

In an FBR, the loss of sodium coolant will lead to overheating, but as there would not be the same effect of losing the moderator in a conventional reactor the reaction will proceed producing more heat. In addition the sodium coolant could burn or react with water (such as the water contained in concrete) causing further fire damage. Together with the high power density of an FBR core and the presence of large amounts of plutonium, still one of the most toxic substances known, this makes for potentially a much more serious situation than a leak in a PWR.

If a reactor were uncontrolled in one of these situations and became supercritical, it would continue to heat up until **meltdown** occurred. At this point the control rod and coolant systems are destroyed and the reaction in the molten core could not be stopped, slowed or contained – a major nuclear disaster such as Chernobyl would occur (see Box 7.4).

Box 7.4

Lessons from Chernobyl

On the morning of the 28 April 1986 routine Swedish radiation monitoring found elevated radiation levels containing anomalous levels of certain radioisotopes. They concluded that it was not coming from any Swedish source, but from a Soviet reactor hundreds of kilometres away, blown on the wind. This was to be the first indication the west had of the worst nuclear accident in history, that had occurred two days previously at Chernobyl.

The Chernobyl disaster occurred as a result of a complex sequence of events (described by Scott and Johnson 1997) during the course of a routine test of reactor operation that resulted in almost all the control rods being removed from the core, against recommended operating guidelines. The RBMK reactor design has a water coolant and graphite moderator, and suffers from an abnormally high positive void coefficient. Because of this, together with high xenon-135 levels, an increase in steam in the core coolant caused the reactor to go supercritical, and the control rods were too far removed to reduce the reaction to sub-criticality quickly enough. Operators and safety systems failed to shut down the system, and the uranium fuel commenced to melt. Control now became impossible, and an explosion followed caused by hydrogen gas released from catalytic breakdown of superheated water. This blew the roof off the reactor building and set fire to the graphite moderator rods. The dispersal of the fuel finally reduced the nuclear reaction rate to subcriticality, but fires continued along with the release of radioactive material for ten days.

The radioactive cloud produced spread to the west on the prevailing winds, to be deposited over Scandinavia and north-western Europe. The spread of radionuclides depended upon their particle size, so for instance plutonium sedimented out within 30 km from Chernobyl, while caesium and iodine condensed on to atmospheric aerosols carrying them to the UK and further. Some radionuclides were projected by the explosion into the stratosphere, dispersing them globally. In the immediate aftermath, the UK population was warned against drinking rainwater and fresh milk production in many areas had to be destroyed. Fallout from the cloud depended upon rainfall patterns, with some areas escaping lightly but heavy deposition in others such as Wales and Cumbria in the UK, where rainfall occurred at an inopportune time. The presence of material from Chernobyl can be identified and distinguished from local sources (including Sellafield in Cumbria and Trawsfynneth nuclear station in Wales) by studying the isotopic ratios of elements such as caesium in the environment.

Meanwhile the Soviet authorities had finally made the accident public allowing a more coordinated response. Initial control was virtually a suicide mission, as workers struggling to put out fires and contain the nuclear material were exposed to radiation levels many times higher than any occupational limits. Later work used robots and remote devices to reduce human exposure. The remains of the reactor was encased in a concrete sarcophagus to contain the radioactivity until activity levels had subsided, and surrounded by a 30-km exclusion zone, from which some 90,000 residents were indefinitely evacuated. These arrangements continue to the present day. Although some remedial work has been carried out to reduce the risk of further release from the remains of the plant, the exclusion zone remains and will do so for many years to come.

The Chernobyl disaster was responsible for 237 cases of acute radiation sickness among workers at the plant and those living very close, of whom 31 died (Anspaugh *et al.* 1988). There has also been a large increase in thyroid cancer cases across areas close to Chernobyl since the disaster, many of which could have been prevented if the authorities had distributed iodine tablets more rapidly (Balter 1995). Many other cancers including leukaemia have occurred, appearing in a small percentage of those exposed often many years later. While estimates vary widely, one statistical study predicts that 16,000 excess deaths worldwide will eventually be caused by the disaster, spread over the 3 billion people exposed mostly to low levels of radiation (Botkin and Keller 2000: 379).

The radioactive fallout elsewhere has been generally at low levels but has been persistent and proved damaging economically. For instance, in upland areas acidic soils keep caesium in solution in the soil water, such that it is absorbed by grass, sheep eat the grass, and slowly expel the radionuclides via their droppings, when they are once more absorbed by the grass. This cycling led to many parts of both Cumbria and Wales being declared unsafe for milk or lamb production until the late 1990s – sheep must be grazed in lowland areas to reduce their caesium levels before selling.

The root cause of the accident has been variously attributed to a poorly trained workforce, bad reactor design and inadequate safety systems. The culture of secrecy and a lack of awareness of safety issues among staff undoubtedly contributed to the seriousness of the accident. Operators broke numerous safety procedures during the sequence of events leading to the accident, but operator error is not a valid explanation, as systems should be designed to make them failsafe. One lesson that may have been learnt is that, without an effective safety culture, where the technology allows operators to cut corners they may well do so, despite safety regulations.

This was not an old or badly-maintained plant, but ranked amongst the most reliable in the USSR and had only been running for two years. However, operator error together with RBMK design faults, namely a lack of alternative control or shutdown systems and a high positive void coefficient, led to supercriticality. The RBMK is also thought to suffer from insufficient delayed neutrons, making control much more difficult. It is for this reason that western experts agree that a similar accident is unlikely to occur in a PWR or AGR type reactor. However the RBMK design is still in use in several nuclear stations in eastern Europe and the former Soviet Union, and in fact the other three reactors at Chernobyl continued to operate after the accident, as they are vital to the Ukraine's power supply.

More common are more minor incidents, involving minor anomalies, deviation from agreed practice or safety checks not carried out, or failure of safety systems when checked. The International Atomic Energy Authority (IAEA) categorises all nuclear incidents according to their seriousness on the International Nuclear Event Scale, which runs from level 1 to 7, and all incidents are monitored. Other safety monitoring includes recording the number of unplanned automatic trips, when the automatic system detects some anomaly and shuts down the reactor, which happens typically once a year.

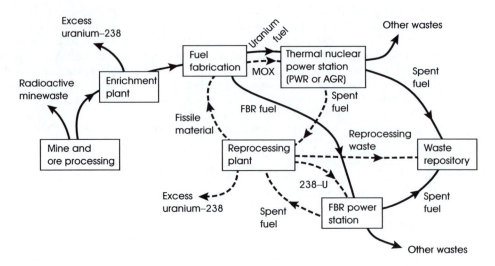

Figure 7.5 *The nuclear fuel 'cycle'. (Dashed arrows represent flows due to reprocessing.)*

The nuclear fuel 'cycle' and reprocessing

The nuclear fuel 'cycle' consists of mining, processing of ore, enrichment, use, fuel and waste transport, reprocessing, and disposal. Each stage produces some waste which contains various radioactive materials (Figure 7.5).

From the mine, uranium ore is refined to 'yellowcake' or U_3O_8, used in fuel rod production, and also produces residues in the form of tailings, containing many different decay products. Minewaste and tailings can produce radioactive radon gas and dust that may be inhaled, and both solid and waterborne radioactive contamination.

The enrichment process increases the proportion of ^{235}U to ^{238}U from the 0.7 per cent found in natural uranium to around 3 per cent for use in reactors. Enrichment is a technically complicated process, as the two isotopes cannot be divided chemically – they must be separated by some process relying on their atomic mass, which may be by atomic centrifuge, gaseous diffusion, mass spectrometry or laser based processes. The fuel rods are then fabricated from uranium oxide, UO_2 (or uranium metal in the case of Magnox reactors).

After use, fuel rods containing the highly radioactive spent fuel may be reprocessed. Spent fuel contains mainly non-fissile ^{238}U, together with unreacted ^{235}U, a range of fission daughter products, and smaller amounts of ^{233}U, plutonium, and other heavy radioactive isotopes produced by neutron absorption (actinides). Reprocessing is the extraction of uranium-238 and fissile isotopes including unreacted ^{235}U, ^{233}U and plutonium. The remaining waste, consisting mainly of fission daughter products, will still require safe disposal.

Uranium-238 (termed depleted uranium, or DU) in theory could be used as fuel in an FBR, and is also used in arms production. The fissile isotopes together with plutonium can be fabricated into 'mixed oxide' fuel or MOX. MOX can be used at a conventional thermal nuclear reactor in place of newly mined uranium, but its use is not straightforward or widespread, as it does require changes to reactor operation, and in fact most reprocessed plutonium and ^{238}U is not used but stored. The British plant THORP (Thermal Oxide Reprocessing Plant) in Cumbria is designed to take spent nuclear fuel from around the world for reprocessing, returning the reusable MOX fuel and waste generated to the country of origin (including Japan and Germany). Technically, reprocessing is achieved by a series of chemical solvent extractions, but is made more complex by the need to prevent overheating from the radioactivity and to prevent the wastes becoming critical.

Reprocessing is controversial and only occurs in a few countries, including the UK, Russia and France; it is no longer policy in the USA, Canada, Sweden and others to reprocess, being seen as more advantageous to dispose of spent fuel directly. It is claimed by its proponents to reduce nuclear waste by removing and reusing the most active part of the spent fuel, and to reduce the need for mining fresh uranium with the concomitant radioactive emissions. If the use of nuclear power were to increase significantly worldwide, reprocessing would become more important in prolonging the lifetime of natural uranium resources. But it also produces more routine emissions, requires highly radioactive spent fuel to be transported around the world, and increases the risk of an accident or other accidental release of radiation. Reprocessing has historical links with the arms industry: the original reason for extracting plutonium was for weapons use, not for civilian reactor fuel. Producing concentrated plutonium may have security risks and present a terrorist target. The economics of reprocessing depends principally on the prices of natural uranium and of the arms 'by-product' of plutonium – prices of both of these have fallen dramatically since the 1970s and the end of the cold war, making reprocessing an uneconomic option.

Whether or not spent fuel is reprocessed, the final stage in the 'cycle' is waste disposal, discussed on pp. 272–7.

Radioactive discharges

In addition to solid waste, discussed on pp. 272–7, a nuclear installation will make radioactive discharges to air and sea on a routine basis. These discharges are regulated to protect the public, on the basis of the estimated dose to the critical population and to the population as a whole. In general the routine discharges are very low, presenting doses to even the most critical groups that are much less than background radiation, although unauthorised discharges do occur.

Discharges to the atmosphere include the gases tritium, (^{3}H), argon-41, krypton-85 and carbon-14 (as CO_2). Of these carbon is the most significant as it can be taken up by plants or animals. The isotopes of most concern discharged to the marine environment are those from reprocessing, including caesium-137, plutonium, americium and technetium.

The **emissions inventory** describes the amount of each radioisotope included in a discharge, which will be monitored. Emission inventories are regulated by discharge limits for the total amount of radioactivity emitted and for specific isotopes. As radioactive decay occurs, the inventory will continue to change for many years after the discharge. These changes result in increases in some isotopes in the environment which may alter the pathways available to the material and its ecotoxicology.

The overall radioactivity of the discharge will decline over time as it decays, as the most active isotopes decay most rapidly. The amount of each radioisotope remaining in the environment depends upon the cumulative emissions over time combined with its decay rate (or half-life). For instance, the reprocessing plant at Sellafield in the UK has been making authorised marine discharges since 1952 at progressively lower levels, into the Irish Sea. Within this relatively closed water body the levels accumulate, dependent upon ocean currents and sedimentation processes. Cumulative quantities of the shorter lived isotopes, caesium-137 (half-life: 30 years) and plutonium-241 (half-life:14 years), are now reducing, as the amounts already discharged have decayed and the emissions are now greatly reduced. However the plutonium-241 decays into americium-241, which is now increasing in concentration as a result. Plutonium-240 and -239 have half-lives of around 24,000 years and so cumulative amounts are not reducing, despite emissions having ceased since 1987.

Decommissioning of nuclear facilities

Decommissioning refers to the dismantling of a nuclear station at the end of its life. In the UK, the older Magnox stations are now around 40 years old, and the decommissioning process has started at several. These stations are finding it harder to meet safety standards and cannot be run economically, but decommissioning is likely to be expensive.

After several decades of operation, much of the fabric of a nuclear reactor becomes mildly radioactive. This is due to neutron activation. While in operation, the nuclear core produces neutrons that penetrate the components within the reactor core, the reactor vessel, and parts of the building. These neutrons are absorbed by nuclei in everyday building materials such as concrete, steel, or carbon, turning them into radioisotopes. It is these activated materials that are the main source of decommissioning wastes.

The first stage in decommissioning is removal of the spent fuel rods, containing at least 97 per cent of the radioactivity of the entire plant, which will be stored and contained elsewhere until eventual disposal. After the fuel rods have been removed, the remaining radioactive components will be contained in the existing reactor building which would be reduced in size, and the rest of the site will be kept secure, monitored and left in the case of Magnox reactors for 135 years, allowing levels of radioactivity to subside. This is Magnox Electric plc's 'Safestore' policy (Magnox Electric plc 1996). After this time, remaining radioactive components are removed for disposal elsewhere, the building would be demolished, and the whole site cleared. The wastes produced all need specialised treatment, but the site itself will be decontaminated (where necessary by removal of soil for disposal), and made suitable for other purposes.

The long delay before final site clearance may seem ill-advised, however there are sound technical arguments for such a delay. Immediately after it has ceased operation, the core of the nuclear plant will contain very high levels of radioactivity and high concentrations of particular products of the nuclear reactions. After a given length of time, activity reduces as some of these products will have naturally decayed, producing α or β radiation that can be easily contained. Some of the isotopes have half-lives of the order of thousands or millions of years, so any reasonable delay would make no difference. Others with very short half-lives decay in a matter of days or months, and so a long delay is not required. However isotopes with half-lives measured in years or decades will have decayed significantly after this time. This is illustrated in Figure 7.6. In the case of a Magnox reactor, after the fuel has been removed, the main isotopes of concern are those produced by neutron activation in trace elements in the carbon steel of the reactor vessel. In the short term the dose rate is dominated by γ-radiation from cobalt-60 (half-life: 5.27 years). Once the cobalt has mostly decayed, niobium-94

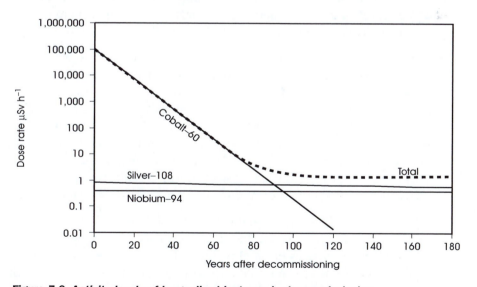

Figure 7.6 *Activity levels of longer-lived isotopes in decommissioning.*

(20,000 years) and silver-108 (418 years) become the dominant sources, which decay more slowly, with the period of 135 years representing the point at which cobalt-60 activity becomes less than silver-108 and niobium-94 (Woollam 1998).

Early decommissioning would require very expensive remote control techniques and present higher risks to workers involved – delaying decommissioning until radioactivity has subsided to more tolerable levels means that the work is both safer and cheaper when it is accomplished. Delay has favourable financial implications, in that it appears far cheaper because the future cost can be discounted. Also if any unforeseen problems are encountered they will have to be solved and paid for by future generations rather than the current nuclear industry. These political and economic considerations may be of equal importance as the technical factors in the decision to delay decommissioning. However delay is not necessary – three reactors in the USA have been decommissioned completely without delay.

Nuclear waste

Nuclear stations produce waste throughout their operating lives as well as at decommissioning. Other sources of radioactive waste include hospitals, pharmaceutical companies, nuclear submarines and research labs. Nuclear waste is generally divided into three classes, determining the disposal method:

- low-level waste (LLW), including everything from workers' clothes to building rubble, with activity less than 2 mGy h^{-1};
- intermediate-level wastes (ILW) – parts of the reactor, fuel cladding, and products from reprocessing and enrichment, activity from 2–20 mGy h^{-1}; and
- high-level waste (HLW), principally spent fuel rods, which produce heat and contain fissile material, activity over 20 mGy h^{-1}.

For any nuclear waste, activity level will decline over time as radioactive decay occurs, hence storage and delay of final disposal until activity is lower is a component of the waste management strategy. Low-level waste is large in volume (millions of cubic metres), but low in activity and can be disposed of by a principle of dilute-and-disperse. Liquid and gaseous waste discharges are made by pipe out to sea and into the air while solid wastes are containerised in ground-level concrete lined vaults at specially designated sites, such as Drigg in Cumbria in the UK. These wastes contain low levels of contamination and these methods are licensed and regulated in order to present no unnecessary risks to the public.

Intermediate-level wastes contain higher levels of radioactivity and long-lived isotopes, and need to be concentrated and contained in some way. High-level waste presents the most serious disposal problem. The volumes produced are relatively small because the energy produced per unit volume of nuclear fuel is so high compared with other fuels. ILW and HLW are often now considered together, as

Plate 7.1 *Low-level solid waste store at Drigg in Cumbria.*

environmental standards have become stricter. These wastes produce radiation and so require containment and shielding.

The spent fuel is many millions of times more radioactive than the uranium was before it was used in the reactor, because it contains highly radioactive fission products. For instance, enriched PWR fuel has total activity of around 1.6×10^{10} Bq t^{-1}, compared with 5.3×10^{18} Bq t^{-1} for spent fuel, after six years in the reactor (Scott and Johnson 1997). While the daughter products of the fission reaction are mainly fairly short-lived, some of the actinides and plutonium have lifetimes of tens of thousands of years. Activity of the waste initially will decrease significantly as shorter-lived isotopes decay (including strontium-90 and caesium-137 with half-lives around 30 years), until the point after several decades when longer-lived isotopes including plutonium-241 start to dominate.

Plate 7.2 *Vitrified high-level waste store at Sellafield, UK.*

As HLW decays it produces energy in the form of heat, so it requires cooling. Initially, fuel rods are held in cooling ponds, which become gradually more radioactive as the water dissolves small quantities of any leaked isotopes and surface contamination, and the concrete lining becomes contaminated with fission products. After reprocessing (if this is done), the remaining waste is mainly fission products, held as a liquid. After several years of storage, once activity has declined, the next step is vitrification, which involves sealing the wastes in glass blocks to make handling easier. Alternatively the waste may be set in concrete or other material to solidify it.

In the long term, there are a number of possible disposal options each with its problems. Neutron activation, which involves bombardment with neutrons in a nuclear reactor, to convert the most harmful and long-lived radionuclides into safer or short-lived ones, is undergoing research. This method if developed could provide an effective method of treating high-level waste without the need for long-term containment, but is likely to be expensive. Extra-terrestrial disposal, that is transporting the waste into space or into the Sun, would be very expensive and involve risks in the launching process as well as having certain ethical problems. Deep geological disposal either on land, on the sea bed or under the permanent ice sheets of the poles must demonstrate that waste will be safely contained over a long timescale.

Of these the current preferred option for both economic and technical reasons is disposal in a deep, land-based repository. This entails deep burying, contained by a combination of artificial and natural barriers, under suitable geological conditions, to reduce dispersion of the waste until activity has subsided to safe levels – at least 10,000 years (see Box 7.5). Due to the long timescale, the disposal site in the long term must not require any active management or institutional control, but must be intrinsically (or passively) safe.

No containment system could permanently isolate the wastes, given the many unpredictable geological factors. The aim is to contain the wastes both physically and chemically such that leakage is small, slow and only in the very, very distant future. Overall the expected rates of escape can be modelled for all radioisotopes in the waste inventory, taking into account such factors as corrosion rates, risk of failure of containers, groundwater movements and diffusion. One such modelling study showed that over a timescale of 100,000 years, less than 0.001 per cent of most radioisotopes and 0.03 per cent of iodine-131 would reach the surface groundwater, under 'worst-case' assumptions on groundwater movement (Dormuth *et al.* 1995). However there are many unknowns in such theoretical estimates – notably the effect of fissuring and colloidal movement.

Could nuclear power help reduce climate change?

Nuclear power does provide an important source of electricity with low emissions of CO_2 and as such has been hailed by some as the solution to global warming. However its benefits are limited. First, nuclear power is only useful for electricity production, and yet from 60 to 80 per cent of energy is used for heat and transport, for which electricity is less suitable. Nuclear stations must be run 24 hours a day to recoup their capital costs, and technically take a very long time to be brought up to power as rapid thermal expansion must be avoided. This means they are limited to provision of low-value baseload electricity, perhaps 30–40 per cent of electricity demand, unless a cheap means of storage is invented – or perhaps off-peak production could be used in domestic storage heaters and to re-fuel electric vehicles.

Box 7.5

Designing a deep repository for nuclear waste

While research in different countries has produced plans for repositories with slight differences in materials or geological conditions available, the fundamental principles are similar. The multiple barrier approach involves sealing the waste inside barriers including the waste container, repository seals and surrounding geological environment. The waste in the form of fuel rods or vitrified reprocessing waste would be sealed in containers made from copper or titanium, to resist corrosion. These would be placed in tunnels bored into rock at a depth of between 500 and 1,000 m. Each container would be surrounded by a clay buffer, and the whole sealed with backfill and cement. All tunnels and shafts would later be sealed to prevent escapes of material from within, or intrusion by people from above. The clay serves to restrict water movement, limiting both corrosion of the container and diffusion of escaped material.

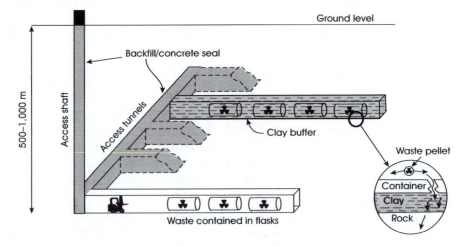

Figure 7.7 *A deep repository for nuclear waste – the multi-barrier concept.*

The barriers provide containment over different timescales according to the half-lives of material contained in the waste. The waste form itself would serve to contain most of the short-lived isotopes, which would not be expected to escape from the fuel pellets or vitrified waste within their lifetimes. The containers would be expected to last from 500 to tens of thousands of years, by which time a large proportion of the waste would have decayed leaving a handful of radioisotopes with very long half-lives, and overall activity reduced by a factor of around 100,000 (Dormuth *et al.* 1995). Movement through the impermeable clay buffer will be largely by diffusion, and hence slow, but fastest for the lighter and more soluble elements such as iodine-129 and carbon-14. Factors such as pH, temperature and the formation of colloids in the groundwater are critical in determining transport rates of the various radionuclides. Any one isotope must be effectively contained by at least one of these barriers, such that the overall effect is to contain all isotopes present in the waste.

The hydrogeology of the proposed site is of great importance, with groundwater movements dependent on rock type, topology, and the freshwater/saline interface. To reduce rates of escape from the repository it would be located in an impermeable rock type such as granite, clay, salt or basalt. Ideally there would be very slow groundwater movement, or with any movement towards the sea, minimising upwards movements or contamination of freshwater aquifers. In designing a deep repository it is important that the surrounding rock is not fissured, which could lead to relatively rapid groundwater movements, potentially both increasing corrosion of the containers and movement of any escaped soluble material into surface groundwater. In addition the repository must be in a geologically stable area, as even a minor earthquake could cause fracturing and damage to the containment systems.

The location of suitable sites has proved difficult, for political as well as technical reasons (see Michie (1998) for a review of the UK experience). Accurate characterisation of rocks at these depths is difficult and expensive, and of course plans for a repository will provoke a classic 'not in my back yard' response from those living locally. It has been suggested that an international repository might be more acceptable, as it could be built in the most geologically suitable area, and a small number of very large repositories may be safer than a large number, one in each country using nuclear power. However this would present further risks and costs from transporting nuclear waste as well as political issues in ensuring international safety.

If the issues of safety, waste and decommissioning could be satisfactorily resolved, and nuclear power were to be adopted worldwide on a large scale, depletion of uranium reserves would more rapidly become an important issue. This could only be countered by successful development of fast-breeder reactors, with their associated problems of cost, safety and plutonium proliferation.

Nuclear power is not cheap, partly as a result of the expense of adequate safety and waste management measures, and partly because of intrinsically high capital costs. Nevertheless, as a relatively short-term measure, nuclear power could have a role in reducing fossil fuel use, at a cost. The decision to be made is whether the economic and environmental costs of nuclear power are justified, or whether the reductions could better be made by a combination of energy conservation and use of renewable forms of energy.

Fusion reactions

Fusion reactions occur when two small, light nuclei meet and fuse into one. This can occur because the mass of the product is less than the sum of the two nuclei before fusing, with the difference being the increased binding energy. A fusion reaction is similar to that occurring naturally in the Sun. Hydrogen or its isotopes, deuterium and tritium, combine to form helium, releasing large amounts of energy.

For instance:

$$\,^{2}_{1} + \,^{3}_{1}H \rightarrow \,^{4}_{2}He + \,^{1}_{0}n \tag{7}$$

deuterium + tritium → helium + a neutron

This reaction is generally called the D–T reaction, and is the basis for most designs of fusion reactor in research into fusion power.

In an H-bomb, the plutonium core is surrounded by hydrogen or deuterium. When the fissile plutonium explodes, the extreme shock forces are sufficient to make the hydrogen undergo fusion, releasing yet more energy – and so creating an even more massive explosion.

Energy in a fusion reaction

Another possible fusion reaction is the 'D–D' reaction (Deuterium–Deuterium) shown in (8), which forms tritium. The tritium may then react with another deuterium in a D–T reaction as in (7) to give helium.

$$2\,^{2}_{1}H \rightarrow \,^{3}_{1}H + \,^{1}_{1}H \tag{8}$$

The energy released comes from the difference in masses between left- and right-hand sides, as for a fission reaction. Given the following masses, in atomic mass units, the energy produced per nucleon pair can be calculated.

Mass of hydrogen:	1.00783 u
Mass of deuterium:	2.01410 u
Mass of tritium:	3.01605 u

Mass on left-hand side $= 2 \times 2.01410 \qquad = 4.02820$ u

Mass on right-hand side $= 3.01605 + 1.00783 = 4.02388$ u

Difference $= 0.00432$ u

This can be converted into more familiar units, as 1 amu $= 1.5 \times 10^{-13}$ J

Energy per nucleon pair $= 6.5 \times 10^{-13}$ J

In 1 kg of deuterium there are 6.2×10^{26} atoms, i.e. 3.1×10^{26} reactions between pairs

Energy per kg $= 3.1 \times 10^{26} \times 6.5 \times 10^{-13}$ J

$= 2.0 \times 10^{14}$ J kg^{-1} or 200 TJ kg^{-1}

This compares with around 29MJ kg^{-1} for coal, nearly seven million times less.

Power from nuclear fusion?

Potentially fusion could provide huge amounts of pollution-free energy, using hydrogen from water as its only raw material, and producing the inert, non-radioactive gas helium as its only waste product. The resources are virtually unlimited as the energy produced per kilogram is so high. Deuterium could be extracted from sea water, and tritium (if used) could be produced from a fusion reaction between hydrogen and lithium, which is abundant. Unlike fission reactions, there are not large numbers of radioactive decay products. The neutrons produced would be easily swept up by absorbing into water, producing more deuterium. Tritium would be produced, but could be used in further fusion reactions.

Technically, harnessing the energy from this reaction as a power source is not easy. To make the reaction occur, the fuel must be kept at incredibly high temperatures of around 100 million degrees. At this temperature the reactants must be in the form of a plasma, where atoms are stripped of their protective electrons, to allow the nuclei to get close enough for fusion to occur. The fuel density must also be kept very high, and these conditions must be maintained for long enough for significant energy to be released.

Maintaining these conditions defies use of any solid reactor vessel, so research uses ideas such as magnetic confinement or laser reactions. In the first, plasma is controlled by a powerful magnetic field to contain it, generally in a torus or doughnut-shaped stream. In the latter, a ball of fuel is subjected to intense input of energy from a laser to make it suddenly react. The problem is that unless the reaction can be sustained for a long time, you have to put more energy into the laser or the magnetic confinement system than you get out of the reaction. Although fusion has been made to work on a small experimental scale, no one has ever managed to get out more energy than they put in. As research continues to make progress, there are still hopes that this situation will change and useful power could be produced.

Compared with nuclear fission reactors, a working fusion reactor would have power density many times higher, with consequent coolant problems in extracting the energy as fast as it was produced. There would be safety and therefore cost implications involved in working with such high energy densities, such that many believe even if it were technically possible fusion could never be economically viable.

Summary

- In a nuclear reaction, large amounts of energy may be released as a result in changes in the binding energy between nucleons.

- Ionising radiation comes in many forms, characterised by having sufficient energy to ionise other atoms or molecules leading to chemical changes and biological damage.

- Radioactive isotopes undergo radioactive decay by emission of α and β radiation.

- Decay occurs after a random time interval characterised by the isotope's half-life, which may be from seconds to billions of years and is not alterable by any common means.

- Fission reactions involve heavy nuclei such as ^{235}U splitting to form two lighter isotopes.

- Nuclear fission involving thermal neutrons, controlled by neutron absorption and moderation, provides the power source in nuclear stations, with the detailed configuration affecting both waste arisings and the risk of accidents.

- Throughout the nuclear fuel cycle, including mining, enrichment, power production, reprocessing and waste disposal, operations are constrained by environmental constraints on routine or accidental release of radioactive material.

- Nuclear fusion provides the potential for a virtually limitless energy source with minimal environmental impacts, but major technical and economic hurdles need to be overcome.

Questions

1 Ordinary nitrogen has atomic number 7, atomic mass 14. How many neutrons, protons and electrons does it have?

2 Uranium-235 has atomic number 92, and atomic mass 235.043924. What is its binding energy?

3 Complete the following reactions, and state what type of nuclear reaction each is (radioactive decay by α or β radiation; nuclear fusion or nuclear fission).

(a) $^{131}_{53}\text{I} \rightarrow ^{127}_{51}\text{Sb} + ?$

(b) $^{60}_{27}\text{Co} \rightarrow ^{60}_{28}\text{Ni} + ?$

(c) $^{233}_{92}\text{U} + ^{1}_{0}\text{n} \rightarrow ^{137}_{54}\text{Xe} + ^{90}_{38}\text{Sr} + ?$

(d) $^{238}_{92}\text{U} + ^{1}_{0}\text{n} \rightarrow ^{239}_{92}\text{U} \rightarrow ^{239}_{93}\text{Np} + ? \rightarrow ^{239}_{94}\text{Pu} + ?$

(e) $^{239}_{94}\text{Pu} + ^{1}_{0}\text{n} \rightarrow ^{137}_{55}\text{Cs} + ^{100}_{46}\text{Pd} + 3? + 7?$

(f) $2\,^{2}_{1}\text{H} \rightarrow ^{3}_{1}\text{H} + ?$

4 A nuclear worker is exposed to a thermal neutron beam and receives a dose of 20 millirads per hour for an exposure time of 5 min. What is his/her dose, in millirads, and equivalent dose in millirems? If the annual dose limit is 5,000 mrems what fraction of it has he/she received?

5 An ancient text is found to have carbon-14 level of 0.006 μCi (micro-curies), compared with 0.007 μCi in the atmosphere. How old is it?

6 A PWR in a year produces spent fuel that contains plutonium-241 with an activity of 5×10^{12} Bq. How long will it take for this level to reduce to 5×10^{11} Bq?

Answers to numerical parts

2 1.9145 amu or 1782 MeV

4 1.67 mrad or 3.33 mrem; 0.067% of annual dose limit

5 1,240 years

6 79,750 years

Further reading

Energy, Society and Environment. D. Elliot. Routledge, London, 1997. Chapter 5 contains a useful, concise discussion of nuclear power issues.

Radioactivity in the Environment: Sources, Distribution and Surveillance. R. L. Kathren. Harwood Academic Publishers, Amsterdam, 1984. Fairly comprehensive coverage of issues and analysis of radioactivity.

Science Matters: Nuclear Power (2nd edn). M. Scott and D. Johnson. The Open University, Milton Keynes, 1997. A comprehensible introduction to nuclear power, including technologies, risk and social issues.

Environmental Science: Earth as a Living Planet (3rd edn). D. B. Botkin and E. A. Keller. John Wiley & Sons, New York, 2000. Chapter 18 is a simple introduction to nuclear power in a glossy format; also contains information on radon in homes.

Appendix A: Mathematical hints

This appendix aims to outline some of the basic mathematical procedures for those with little mathematics background, to enable simple calculations to be performed and act as a reminder of basic algebra.

Scientific notation

In physics, numbers are often written in scientific notation, for instance 1.23×10^5 (1.23 times ten to the five). This means 123,000: the number with the decimal point shifted five places to the right. The number 5 is termed the exponent. Small numbers are written with a negative exponent, for instance 5.43×10^{-2} means 0.0543.

When multiplying these numbers, you can add the exponents. When dividing, you can subtract them. This makes it easy to estimate things in your head, e.g.:

$$1 \times 10^4 \times 3 \times 10^{-2} = 3 \times 10^2$$

$$4 \times 10^6 \div 2 \times 10^3 = 2 \times 10^3$$

On a calculator, these numbers are entered using the EXP key, and are displayed as 1.23 E 5 or 5.43 E -2. The E stands for exponent. (On some calculators there is no E, just a space.)

Conventionally in scientific notation the number is given with one figure to the left of the decimal point – e.g. 1.23×10^5 rather than 12.3×10^4, even though mathematically these are the same number. However you may also see numbers given with powers divisible by three (0.123×10^6 or 123×10^3), as this corresponds with the prefixes on the unit of kilo, mega and so on (Table B1).

Few quantities are known to within more than a few decimal places, and errors are increased when numbers are multiplied or otherwise combined. Always avoid spurious accuracy – write 1.2×10^{24} rather than $1.239562837 \times 10^{24}$, even if your calculator tells you the latter answer.

Scientists talk of 'order of magnitude' estimates, which strictly speaking means within one power of ten, i.e. between a tenth and ten times the true answer. While this may not be useful for some things – for instance your salary – in other cases it may be all you need to know.

Mathematical functions

Table A1 lists mathematical functions that are commonly used:

Table A1 *Mathematical functions*

Function	Written	Notes
Powers	a^x, a^{-x}	$a^x = 1/a^{-x}$ and vice versa
Inverse	$1/x$	Equivalent to x^{-1}
Exponential	e^x	Where e = the constant 2.718 . . .
Logarithm	$\log (x)$	If $y = \log (x)$ then $10^y = x$
Natural logs	$\ln (x)$ or $\log_e (x)$	If $y = \ln (x)$ then $e^y = x$ where e = the constant 2.718 . . .
Roots	$\sqrt{x}$, $\sqrt[3]{x}$	If $y = \sqrt[n]{x}$, $x = y^n$ $\sqrt[n]{x}$ can also be written $x^{1/n}$ $1/(\sqrt[n]{x})$ can be written $x^{-1/n}$

Order of calculation

It is important to calculate a formula in the correct order, not necessarily the order it is written – this order is:

1 anything in brackets
2 mathematical functions
3 ×, ÷ (or /, ___)
4 +, −

N.B. The times sign is usually not used but implied between any two adjacent terms. The 'over' symbol means the same as divide, but calculate everything on the top and everything on the bottom before dividing, as if each were in brackets. Thus

$$\frac{A + B}{CD}$$ means the same as $(A + B) \div (C \times D)$.

If these rules are not applied the answer will be wrong. Try these:

$$2 + 6 \div 3 \times 4 - 1 = 9$$

$$(2 + 6) \div (3 \times 4) - 1 = -0.333$$

$$\frac{2 + 6}{3 \times 4 - 1} = 0.727 \ldots$$

Note also that, for instance, $(3 + 4)^2$ is not the same as $3 + 4^2$, and similarly $\log (2 \times 10^3)$ is not the same as $\log (2) \times 10^3$, and so on.

Rearranging equations

When you rearrange an equation, you are either multiplying or dividing each side by the same thing, or adding/subtracting something from both sides, or using a function on both sides. For instance, using Ohm's law (Equation (1.17)) as an example:

$$V = IR$$

Divide by R to find I: $\quad V/R = I$

Another example, rearranging the first equation of motion (Equation (1.2)) to calculate acceleration:

$$v = u + at$$

Subtract u $\qquad v - u = at$

and divide by t: $\quad (v - u)/t = a$

Functions can be reversed by using the inverse function, for instance the inverse of a square is a square root. For instance, rearranging Equation (1.4) to find the initial speed u:

$$v^2 = u^2 + 2as$$

Subtract ($2as$): $\quad v^2 - 2as = u^2$

Take square root: $\sqrt{(v^2 - 2as)} = u$

Likewise, a logarithm is the inverse of an exponential or 10^x function. For instance in Equation (6.3) a decibel is defined as:

$$L = 10 \log (I/I_0)$$

Rearranging to find the sound intensity:

Divide by 10: $\qquad L/10 = \log(I/I_0)$

Apply 10^x: $\qquad 10^{(L/10)} = I/I_0$

Multiply by I_0: $\quad I_0(10^{(L/10)}) = I$

Similarly, for radioactive decay described by an exponential function in Equation (7.1):

$$N/N_0 = e^{(-0.693\, T/T_{1/2})}$$

This can be rearranged by taking the natural log of both sides:

$$\ln (N/N_0) = -0.693\, T/T_{1/2}$$

Now multiply by $T_{1/2}$ and divide by 0.693 to get the age T:

$$T = \frac{T_{1/2} \ln (N/N_0)}{-0.693}$$

Of course not all equations can be simply rearranged like this. For instance to find the time t from the second equation of motion, $s = ut + \frac{1}{2}at^2$ (Equation (1.3)), if u is not zero, requires you to solve this quadratic equation, which may yield two possible solutions or none at all, and so is not quite so straightforward.

There are some special rules for exponents and logs. When the two powers are multiplied, this is equivalent to adding their exponents; similarly division can be done by subtraction of exponents. Conversely, the sum of two logs (or natural logs) is equal to the log of the product of their arguments. For example:

$$10^3 \times 10^5 = 10^8$$

$$\log (3) + \log (10) = \log (3 \times 10)$$

$$e^{(x-y)} = e^x / e^y$$

Adding/subtracting the powers of ten is an easy way of getting an order of magnitude solution to a calculation without using a calculator, and is also useful to crosscheck that your calculated answer is correct.

Proportionality

If a quantity is proportional to another, this means doubling one will double the other, and so on. This may be expressed as an equation by multiplying one quantity by a constant, for instance in Equation (3.1), heat is given by temperature rise, ΔT, multiplied by heat capacity C. C is the constant of proportionality:

$$Q = C \, \Delta T$$

This is a linear relationship, as a graph of Q against ΔT will be a straight line, with gradient C.

The converse is inverse proportionality, where doubling one factor will halve the other. For instance, frequency (f) is inversely proportional to wavelength (λ) of a wave, with the constant being the velocity of the wave, v:

$$f = v/\lambda$$

A quantity may be proportional to several determining factors, which can all be multiplied together. For instance, heat transfer by conduction is proportional to the temperature difference and the area and inversely proportional to the length of conductor, with the constant of proportionality being the conductivity k (Equation (3.5)).

$$Q = \frac{kA(T_2 - T_1)}{l}$$

From this it can immediately be seen that if temperature change or area are doubled, heat transfer will double, while if the distance l doubles, heat transfer will halve.

In many cases, direct proportionality is replaced by being proportional to something squared or cubed – a power relationship. The higher the power, the more one will increase when the other increases. For instance, the power in the wind extracted by a turbine is proportional to the windspeed cubed – without any equations at all it can be seen that doubling windspeed will increase power by a factor of two cubed, i.e. eight times.

Using your calculator

Calculators vary in their details, but these notes may help overcome some teething troubles. The best check is to have some idea of the magnitude of the answer you expect, so that you notice if it is wrong.

To enter a number in scientific notation, enter the number, then press the button labelled EXP or E, then enter the exponent. You can then use them like ordinary numbers. (N.B. *Do not* use x^y or enter the '× 10' – this is done by the exponent key.) If your calculator hasn't got an exponent key, you can calculate the answer without any of the '×10^n's, then add up/subtract exponents by hand (but note this won't work for mathematical functions or logs).

Negative numbers: Depending on your calculator, you may need either to enter [−] then the number, or enter the number, then the [+/−] key (*not* the minus sign). Enter this before the exponent; after it will change the sign of the exponent, not the number, e.g.:

$$- 1.23 \times 10^4 \quad \text{enter as} \quad [-]1.23 \text{ [EXP] } 4$$
$$\text{or} \quad 1.23 \text{ [+/−] [EXP] } 4$$
$$- 2.34 \times 10^{-43} \quad \text{enter as} \quad [-]2.34 \text{ [EXP] } [-]43$$
$$\text{or} \quad 2.34 \text{ [+/−] [EXP] } 43[+/−]$$

Generally it's a good idea to evaluate any mathematical functions in the formula first.

Break large formulas down into bite-sized chunks. Use brackets where necessary if your calculator has them, including when using the 'over' symbol. If your calculator doesn't have brackets, use either the memory or a piece of paper to calculate things in brackets first, then put the formula together.

For mathematical functions, the order of keys pressed is not necessarily the same as what is written down in the formula. To get inverse functions (the small print above the function keys) press [inv] or [shift] before the function key. The exact details vary, but Table A2 lists the typical.

Table A2 *Mathematical functions using a calculator*

Function	Example	Key	Enter	Answer
Power	7^3	x^y	7 [x^y] 3 [=]	343
Log	log (24)	log	[log] 24 [=]	1.38
Natural log	ln (24)	ln or $\log_e$	[ln] 24 [=]	3.178
Exponent	$e^{3.4}$	e^x (inverse ln)	[inv] [ln] 3.4 [=]	29.96
Roots	$\sqrt{9}$	$\sqrt{}$	[$\sqrt{}$] 9 [=]	3
	$\sqrt[5]{17}$	$\sqrt[x]{y}$ or $x^{1/y}$ (inverse x^y)	5 [inv] [x^y] 17 [=]	1.76
Sines and cosines	sin (45°)	sin	[sin] 45 [=]	0.707

Some calculators have a cube root key as well as a square root.

On some calculators you have to press the number first followed by the function key, and you may not need the [=] key – you will have to check the instructions or experiment with your own, e.g.: 24 [log] rather than [log] 24 [=].

The $1/x$ key divides 1 by the number you have just entered. This is useful sometimes if you want to divide by something you have already entered.

e.g. you could calculate

$$\frac{2}{\log (24)} \quad \text{as} \quad 2 \ [\div] \ [\log] \ 24 \ [=]$$

$$\text{or as} \ [\log] \ 24 \ [=] \ [1/x] \ [\times] \ 2 \ [=]$$

and get the same answer (1.449).

Appendix B: Symbols and abbreviations

The basic SI units can all be used with the prefixes given in Table B1 to express very large or very small units, such as MJ (mega joule = 10^6 joules) or nm (nanometre = 10^{-9} m). The prefixes always increase or decrease in powers of three.

Table B1 *Prefixes for units*

Big ones			Little ones		
k	kilo	$\times 10^3$	m	milli	$\times 10^{-3}$
M	mega	$\times 10^6$	μ	micro	$\times 10^{-6}$
G	giga	$\times 10^9$	n	nano	$\times 10^{-9}$
T	tera	$\times 10^{12}$	p	pico	$\times 10^{-12}$
P	peta	$\times 10^{15}$			

Table B2 gives Greek letters most commonly used in physics, and what for, together with a couple of other symbols. Table B3 lists letters and the quantities they are conventionally used to symbolise.

Table B2 *Greek letters*

Letter	Name	Used for
α	alpha	Constant; type of radiation
β	beta	Constant; type of radiation
δ	delta	Small difference
Δ	delta (capital)	Larger difference
ϕ	phi	
γ	gamma	Type of radiation
η	eta	Efficiency
ψ	psi	
κ	kappa	Constant
λ	lambda	Wavelength
μ	mu	Micro; coefficient; viscosity

π	pi	3.142 ...
θ	theta	Temperature; angle
ρ	rho	Density
σ	sigma	Constant
τ	tau	Timespan
ω	omega	Angular velocity
Σ	sigma (capital)	Summation sign
Ω	omega (capital)	Ohms (unit of electrical resistance)
∝	proportionality sign	'Is proportional to'
∞	infinite sign	Infinite

Table B3 *Letters*

Letter	Used for	Letter	Used for
a, A	Acceleration; area; atomic mass	n, N	Number
b, B	Magnetic flux	o	
c, C	Constant; speed of light, coulomb; heat capacity	p, P	Pressure; power
d	Diameter; depth or thickness; differential	q, Q	Electrical charge; quantity of heat
e, E	Exponential constant; electron; energy; Young's modulus; electric field	r, R	Radius; radial distance; electrical resistance
f, F	Frequency; force	s, S	Distance; entropy
g, G	Gravity; universal gravity	t, T	Time; temperature; tesla; time period; torque
h, H	Height; Planck's constant; magnetic field	u	Momentum
i, I	$\sqrt{-1}$; index; moment of inertia; electrical current; intensity	v, V	Velocity; volume; voltage
j, J	Index; Joules	w	Work
k	Constant; coefficient	x	Distance in one direction (axis)
l, L	Litres; length; angular momentum	y	Distance in one direction (axis)
m	Metres; mass	z, Z	Distance in one direction (axis); atomic number

Glossary

absolute temperature temperature measured in Kelvin, a scale starting at absolute zero.

absolute zero temperature of zero Kelvin, indicating that thermal energy of molecules is zero and a lower temperature is impossible.

absorption spectrum range of wavelengths of radiation absorbed by a material.

adiabatic changes in temperature, volume and pressure of a gas that occur without external energy flows.

adiabatic lapse rate normal temperature decrease with height in the atmosphere.

aerofoil asymmetric shape that produces lift from airflow around it.

alpha particle produced by radioactive decay, consists of a helium nucleus.

angular momentum conserved quantity, representing the tendency to continue rotating once doing so.

asthenosphere region of the Earth's mantle that is partially molten and plastic, allowing flow.

atomic mass mass of a nucleus, in atomic mass units, approximately equal to the total number of nucleons.

atomic mass unit unit of mass equal to one-twelfth the mass of a carbon-12 nucleus.

atomic number number of protons in a nucleus.

beta particle produced by radioactive decay, consists of an electron.

binding energy energy that holds atomic nuclei together.

black body radiation electromagnetic radiation produced as a consequence of a body's temperature, at a maximum if from a pure black surface.

carbon cycle flows and sinks of carbon through the atmosphere, biosphere, oceans, soils and rocks.

capillarity action of fluids rising through narrow tubes or pores due to surface tension effects.

Carnot cycle theoretical thermodynamic cycle involving changes in temperature, pressure and volume of a gas in which heat is input and mechanical work output, from which the maximum efficiency of engines can be derived.

centrifugal and centripetal forces forces acting on a rotating body, towards and away from the centre of rotation respectively.

climate space range of ambient temperature and other external factors within which an organism can survive.

combined heat and power (CHP) production of electricity and power simultaneously.

concentration gradient rate of change of concentration with distance.

condensation nucleus solid particle that aids condensation.

conduction movement of heat through the body of a material; flow of electrical current.

convection movement of heat by currents due to density changes brought about by temperature changes.

coulomb unit of electrical charge.

Curie point temperature higher than which a ferromagnetic material will lose its permanent magnetic field.

damping depth distance to which a varying heat source will cause varying temperature when conducted through a medium, typically the depth of diurnal or annual variations in soil temperature.

decibel unit of sound measurement, defined on a log scale relative to a given baseline.

declination direction of the Earth's magnetic field.

diffraction tendency of waves to spread out when passing a sharp edge or small aperture.

diffusion movement of a material due to kinetic energy of its particles.

dimensionless without units, e.g. a ratio.

dispersion spreading out as a result of wind, currents, gravity and other movement.

drag force frictional force between a fluid and an object moving through it, acting to slow down the object.

efficiency the ratio of useful energy output to total energy input.

El Niño an anomalous current system in the southern Pacific, occurring every few years, with impacts upon weather systems across the southern hemisphere.

elasticity extent to which a material may be deformed without suffering irreversible change.

electromagnetic radiation energy transmitted by photons as alternating electrical and magnetic fields, that constitutes light, radiant heat, radio waves, X-rays and other forms.

electron   elementary particle with negative charge, exists in energy levels around atoms, responsible for chemical reactions and electricity.

emission spectrum   range of wavelengths of electromagnetic radiation emitted from a body at a certain temperature.

emissivity   proportion of the maximum (black body) radiation emitted from a warm object, dependent on its colour.

entropy   quantity measuring disorder of a system, related to energy quality, that always increases.

fission, nuclear   reaction involving a large nucleus splitting into two daughter radionuclides, releasing radiation, neutrons and energy.

frequency   the number of wave peaks per second, for any wave including electromagnetic and sound.

fusion, nuclear   reaction involving two light nuclei joining to make one, releasing energy.

gamma radiation   type of high frequency electromagnetic radiation produced by nuclear reactions.

general atmospheric circulation   global system of winds driven by differences in temperature at different latitudes.

geomagnetic anomaly   deviation in the Earth's magnetic field, due to underlying rocks or other magnetic objects.

geostrophic wind   a wind blowing parallel to isobars.

geothermal power   electricity or heat obtained from underground hot rocks or aquifers.

gravitational anomaly   variation in the strength of gravity at a certain point, due to subsurface geological features.

greenhouse effect   term given to the warming of the Earth's surface as a result of the build up of gases such as carbon dioxide in the atmosphere.

groundwater   water that flows or lies underground, both within the pores of rocks or soil and in larger cavities.

half-life   the time taken for half of a radioactive isotope to decay.

heat capacity   quantity of heat needed to raise an object's temperature by one degree.

heat engine   device which converts energy from heat into mechanical form.

heat pump   device which converts a large amount of low temperature, ambient heat into a smaller amount of higher temperature heat.

hydroelectric power (HEP)   electricity generated from moving water.

inclination   slope of the Earth's magnetic field relative to the ground.

insolation   amount of energy from the Sun incident at the Earth's surface.

interference   property of superimposed waves to either cancel out or amplify one another, producing characteristic patterns.

ionisation   freeing electrons from the atoms of a substance, making it chemically reactive.

ionising radiation   electromagnetic radiation of high frequency, with sufficient energy to ionise atoms.

isobar   contour following points of equal pressure (in the atmosphere).

isostasy   the process by which the Earth's crust lifts to compensate for subsurface density, as if it floats on the plastic asthenosphere below.

isothermal   changes taking place in a gas at constant temperature.

isotope   an alternative form of an element with a different number of neutrons in its nucleus.

Kelvin   unit of absolute temperature.

kinetic energy (KE)   energy of a body due to its motion.

laminar flow   flow along smooth, regular, roughly parallel flow lines.

latent heat   extra heat required to bring about change of state from solid to liquid or from liquid to gas.

lithosphere   rigid part of the Earth's surface, consisting of the crust and upper part of the mantle.

loudness level   perceived level of noise, measured in phons.

mesosphere   upper layer in the atmosphere, above the stratosphere, from about 50–90 km height.

moment of inertia   distribution of mass around the axis of a rotating body that determines its rotational properties.

momentum   conserved quantity, representing the tendency to keep moving on a straight line course, dependent on mass and velocity of a body.

neutron   nuclear particle with no charge.

nuclear transmutation   process by which any nucleus can be changed into a different isotope or element, by bombardment with neutrons or other high energy radiation.

nucleon   nuclear particle, i.e. neutron or proton.

ozone layer   region in the stratosphere containing high levels of ozone (O_3), which absorbs ultra-violet radiation.

passive solar   use of solar energy for direct space heating of buildings.

period   time taken for a complete wavelength of a wave to pass a fixed point.

photosynthesis   process by which green plants convert solar energy into chemical form as carbohydrates.

photosynthetically active radiation (PAR)   range of wavelengths used by photosynthesis.

photovoltaic cell (PV)   semiconductor device that converts light into electricity.

permeability   degree to which water can move through a medium such as soil or rock.

plasma   substance at a sufficiently high temperature such that all atoms are completely ionised as a result of their thermal energy.

polarisation   separation of light into waves with electrical fields along different perpendicular axes.

porosity   proportion of a soil or rock consisting of air or water in voids known as pores.

potential energy (PE)   energy of a body due to its position, relative to a gravitational, electromagnetic or nuclear field.

proton   nuclear particle with a positive charge.

Planck's law   states that the energy carried by an electromagnetic wave is directly proportional to its frequency, $E = hf$, with the coefficient h being Planck's constant.

radiation window   range of wavelengths in the absorption spectrum of the atmosphere in which absorption is low and radiation can penetrate.

radioactive decay   process by which a radionuclide can spontaneously decay into another element, emitting an alpha or beta particle.

radioisotope   a radioactive isotope whose nuclei are radionuclides.

radionuclide   an unstable isotope, subject to radioactive decay.

remanent magnetism   magnetic fields created in rocks when they formed.

resonance   property of objects to absorb wave energy at a certain frequency.

Reynolds number   a dimensionless quantity characterising fluid flow, representing the ratio of viscous to inertial forces.

scalar   quantity with a magnitude but no direction.

scattering   tendency of light to spread out in the atmosphere by reflection and diffraction around small particles.

seismic wave   a wave travelling through the Earth, from an earthquake or a major explosion or impact.

smog   air pollution deriving from smoke and fog, often exacerbated by temperature inversion.

solar energy   energy derived from the Sun.

solar wind, solar magnetic flux   stream of charged particles emitted from the Sun that strikes the upper atmosphere.

strain   deformation of a body due to an applied stress.

stratosphere   second layer in the atmosphere, from around 11 to 50 km height.

stress   force exerted per unit area on a body.

supercooled   cooled below freezing point without freezing.

superheated   heated above boiling point without evaporating.

supersaturated   containing more of a dissolved substance than the saturation point.

temperature inversion   atmospheric temperature profile in which warm air overlies cold, which may trap fog and pollution.

terminal velocity   maximum velocity reached when falling, when drag forces equal gravity.

thermal diffusivity   parameter describing temperature changes as a result of varying input of heat.

thermohaline circulation   ocean currents due to differences in temperature and salinity.

troposphere   lowest level of the atmosphere, up to around 11 km, containing weather systems.

turbulent flow   flow containing eddies and unpredictable rapid changes over time, where flow lines are not parallel.

ultrasound   sound with a frequency higher than the audible range.

vector   quantity with magnitude and direction.

viscosity   property of a fluid indicating how thick and 'treacly' it is.

wavelength   length from one peak of a wave to the next, for any wave including electromagnetic and sound.

Young's modulus   measure of elasticity of a material, defined as linear stress divided by linear strain.

Bibliography

Adler, R. (2000) 'Forests turn to dust'. *New Scientist*, 6 May, 7.

Anon. (2000) *The Guardian*, 23 February.

Anspaugh, L. R., Catlin R. J. and Goldman, M. (1988) 'The global impact of the Chernobyl reactor accident'. *Science*, 242, 1513–1518.

Balter, M. (1995) 'Chernobyl's thyroid cancer toll'. *Science*, 270, 1758.

Barnard, G. and Kristofferson, L. (1985) *Agricultural Residues as Fuel in the Third World.* IIED/Beijer Inst. Energy Information Programme, Report no. 4, Earthscan, London.

Barnola, J. M., Raynaud, D., Korotkevich, Y. S. *et al., C.* (1987) 'Vostok ice-core provides 160,000-year record of atmospheric CO_2'. *Nature*, 329, 408–414.

Bennet M. R. and Doyle, P. (1997) *Environmental Geology: Geology and the Human Environment.* John Wiley, New York.

Boeker, E. and van Grondelle, R. (1999) *Environmental Physics.* 2nd edn. John Wiley & Sons, Chichester.

Botkin, D. B. and Keller, E. A. (2000) *Environmental Science: Earth as a Living Planet.* 3rd edn. John Wiley & Sons, New York.

Boyle, G. (ed.) (1996) *Renewable Energy: Power for a Sustainable Future.* Oxford University Press/The Open University, Oxford.

Campbell, G. S. (1977) *An Introduction to Environmental Biophysics.* Heidelberg Science Library, Springer-Verlag, Heidelberg.

Cline, W. (1992) *The Economics of Global Warming.* Institute for International Economics, Washington DC.

Critchfield, H. J. (1983) *General Climatology.* 4th edn. Prentice-Hall, New Jersey.

Davis, Mackenzie L. (2000) *Introduction to Environmental Engineering.* McGraw-Hill, New York.

Day, N. and the UK Childhood Cancer Study investigators (1999) 'Exposure to power-frequency magnetic fields and the risk of childhood cancer'. *The Lancet*, 354 (9194), 1925–1931.

Dormuth, K. W., Gillespie, P. A. and Whitaker, S. H. (1995) 'Disposal of nuclear fuel waste'. In: R. E. Hester and R. M. Harrison (eds), *Issues in Environmental Science and Technology series, vol. 3: Waste Treatment and Disposal.* Royal Society of Chemistry, Cambridge, 115–130.

Elliot, D. (1997) *Energy, Society and Environment.* Routledge, London.

Farmer, J. G., Eades, L. J. and Graham, M. C. (1999) 'The lead content and isotopic composition of British coals and their implications for past and present releases of lead to the UK environment'. *Environmental Geochemistry and Health*, 21, 257–272.

Fews, A. P., Henshaw, D. L., Keitch, P. A., *et al.* (1999a) 'Increased exposure to pollutant aerosols under high voltage power lines'. *Int. J. Radiation Biol.* 75(12), 1505–1522.

Fews, A. P., Henshaw, D. L., Wilding, R. J. *et al.* (1999b) 'Corona ions from powerlines and increased exposure to pollutant aerosols'. *Int. J. Radiation Biol.* 75(12) 1523–1532.

Fleagle, R. G. and Joost, A. B. (1980) *An Introduction to Atmospheric Physics.* International Geophysics Series, vol. 25. Academic Press, London.

Gailitis, A., Lielausis, O., Dement'ev, S. *et al.* (2000) 'Detection of a flow induced magnetic field eigenmode in the Riga dynamo facility'. *Physical Review Letters*, 84(19), 4365–4368.

Green, B. M. R., Lomas, P. R. and O'Riordan, M. C. (1992) *Radon in Dwellings in England*. NRPB report R254. Didcot, Oxon.

Guyot, G. (1998) *Physics of the Environment and Climate*. John Wiley & Sons/Praxis Publishing Ltd, Chichester.

Hall, D. O., Barnard, G. W. and Moss, P. A. (1982) *Biomass for Energy in the Developing Countries*. Pergamon Press, Oxford.

Halliday, D., Resnick, R. and Walker, J. (2000) *Fundamentals of Physics*. John Wiley & Sons, Chichester.

Hill, R., O'Keefe, P. and Snape, C. (1995) *The Future of Energy Use*. Earthscan, London.

Houghton, J. T., Jenkins, G. J. and Ephraums, J. J. (1993) *Climate Change: The IPCC Scientific Assessment*. Cambridge University Press, Cambridge.

Houghton, J. T., Meira Filho, L. G., Callender, B. A. *et al.* (eds) (1995) *IPCC second assessment report: Climate Change. The Science of Climate Change*. Contribution of Working Group 1 to the 2nd assessment of the IPCC. Cambridge University Press, Cambridge (also available online at http://www.ipcc.ch).

IPCC (1995) *Second Assessment Synthesis of Scientific–Technical Information relevant to interpreting Article 2 of the UN Framework Convention on Climate Change*. IPCC, Geneva (also available online at http://www.ipcc.ch).

ISO (1987) *Acoustics – Normal Equal-Loudness Level Contours*. Report ISO 226:1987. International Organization for Standardization, Geneva.

Johanssen, T. B., Kelly, H., Reddy, A. K. N. *et al.* (eds) (1993) *Renewable Energy: Sources for Fuels and Electricity*. Island Press, Washington DC.

Jones, A. M. (1997) *Environmental Biology*. Routledge, London.

Jouzel, J., Lorius, C., Petit, J.-R. *et al.* (1987) 'Vostok ice-core: a continuous isotope temperature record over the last climatic cycle (160,000 years)'. *Nature*, 329, 403–408.

Kathren, R. L. (1984) *Radioactivity in the Environment: Sources, Distribution and Surveillance*. Harwood Academic Publishers, Amsterdam.

Kazmerski, L. L. (1997) 'Photovoltaics: A review of cell and module technologies'. *Renewable and Sustainable Energy Reviews*, 1(1/2), 71–170.

Kiely, G. (1997) *Environmental Engineering*. McGraw-Hill, New York.

Magnox Electric plc (1996) *Environmental Report 1995–1996*. Berkeley, Gloucestershire.

Mannion, A. (1999) *Natural Environmental Change*. Routledge, London.

McKinlay, A. (1997) 'Possible health effects related to the use of radiotelephones'. *Radiological Protection Bulletin* 187, March, 9–16.

Michie, U. McL. (1998) 'Deep geological disposal of radioactive waste: a historical review of the UK experience'. *Interdisciplinary Science Reviews*, 23(3), 242–256.

Monteith, J. L. and Unsworth, M. H. (1990) *Principles of Environmental Physics*. 2nd edn. Edward Arnold, London.

Montgomery, C. (1995) *Environmental Geology*. Wm. C. Brown, Iowa.

Nelkon, M. and Parker, P. (1994) *Advanced Level Physics*. 7th edn. Heinemann Educational Books, London.

New Scientist (2000) 'Barely audible'. Last Word Special, 13 May, 38.

Niebel, G., Hanewinkel, R. and Ferstl, R. (1993) 'Violence and aggression in schools in Schleswig-Holstein'. *Zeitschrift für Pädagogik*, vol. 39, 775.

O'Callaghan, P. W. (1992) *Energy Management*, McGraw Hill, New York.

O'Neill, P. (1998) *Environmental Chemistry*. 3rd edn. Blackie, London.

Open University (1993) *T237 Environmental Control and Public Health: Units 11,12 and 13*. Open University, Milton Keynes.

Ordnance Survey (1995) Map: Landranger 102 Plymouth and Launceston, 1:50,000.

O'Riordan, N. J. and Milloy, C. J. (1995) *Risk Assessment for Methane and Other Gases from the Ground*. CIRIA Report No 152. Construction Industry Research and Information Association.

Pearce, F. (1999) 'Weather warning'. *New Scientist*, 9 October, 36–39.

Price, M. (1985) *Introducing Groundwater*. Chapman & Hall, London.

Rango, A. (1995) 'Effects of climate change on water supplies in mountainous snowmelt regions'. *World Resource Review*, 7(3), 315–325.

Richter, C. F. (1958) *Elementary Seismology*. Freeman, San Francisco.

Ritter, D. F. (1986) *Process Geomorphology*. Wm C. Brown, Iowa.

Rosenfeld, D. (2000) 'Suppression of rain and snow by urban and industrial air pollution'. *Science*, 287, 1793.

Royal Commission on Environmental Pollution (1995) *Eighteenth Report: Transport and the Environment*. Oxford University Press, Oxford.

Schott, J. R. (1997) *Remote Sensing – The Image Chain Approach*. Oxford University Press, Oxford.

Scott, M. and Johnson, D. (1997) *Science Matters: Nuclear Power*. The Open University, Milton Keynes.

Seinfeld, J. and Pyros, S. N. (1998) *Atmospheric Chemistry and Physics: From Air Pollution to Climate Change*. John Wiley, New York.

Seungdoo Park, J., Vohs, J. M. and Gorte, R. J. (2000) 'Direct oxidation of hydrocarbons in a solid-oxide fuel cell'. *Nature*, 404, 265–267.

Sharma, P. V. (1997) *Environmental and Engineering Geophysics*. Cambridge University Press, Cambridge.

Sienkiewicz, Z. (1997) 'ABC of RF'. *Radiological Protection Bulletin* 196, December, 23–28.

Summerhayes, C. P. and Thorpe, S. A. (1996) *Oceanography – An Illustrated Guide*. Manson Publishing, London.

TERES (1994) *The European Renewable Energy Study: Prospects for Renewable Energy in the European Community and Eastern Europe up to 2010*. Main Report. Prepared for Commission of the European Communities, D-G XVII. Office for Official Publications of the E.C., Luxembourg.

Thompson, R. (1998) *Atmospheric Processes and Systems*. Routledge, London.

van Dop, H. (n.d.) *Luchtverontreinig, Bronnen, Verspreiding, Transformatie en Depositie*. KNMI, De Bilt z.j. (Netherlands Meteorological Institute).

Vesilind, P. A., Peirce, J. J. and Weiner, R. (1988) *Environmental Engineering*. 2nd edn. Butterworths, Massachusetts.

Walker, G. (2000) 'The Hole story'. *New Scientist*, 25 March, 24–28.

Watson, K. (1993) *Foundation Science for Engineers*, Macmillan, London.

Woollam, P. B. (1998) 'How will we manage the waste from reactor decommissioning?'. *Interdisciplinary Science Reviews*, 23(3), 281–287.

Index

Entries in *italics* indicate figures; those in **bold** indicate tables; B denotes a text box and P a plate.